Natural Power

Natural Power: The New York Power Authority's Origins and Path to Clean Energy

BY
Rock Brynner

FOREWORD BY
Governor Andrew M. Cuomo

NEW YORK

Cover photograph by the Power Authority of the State of New York
Cover design by www.popshopstudio.com
Interior design by Maria Torres

ISBN: 978-1-944529-36-9

CONTENTS

For a broad collection of photos and portraits from *Natural Power*, go to www.naturalpowergallery.com.

The New York Power Authority is maybe one of our proudest boasts. You almost never hear about the New York Power Authority, but you should. It would do us all a lot of good: you can see government at its very best.

—Mario Cuomo
Governor, New York State

1991

Foreword

New York State has had a unique relationship with generated power since its leading pioneers introduced the world's first electrical grid in Lower Manhattan more than a century ago. Considering how profoundly man-made electricity has transformed almost every aspect of human existence, it is reasonable to argue that the inventions launched here in New York by Edison, Tesla, Westinghouse and others produced the greatest technological revolution since the printing press.

From a twenty-first-century perspective, the development of the power industry is an affirmation of human ingenuity. Thanks to my predecessor, Governor Franklin D. Roosevelt, who created New York's Power Authority, the clean, renewable hydropower pioneered along the Niagara and St. Lawrence rivers belongs to the people of New York and provides us with the energy infrastructure that we need to achieve sustainability.

Now, as you will read, NYPA has sixteen clean hydroelectric and natural gas-fueled power plants and more than 1,400 circuit-miles of transmission lines, whose effective management is crucial to the state's well-being. Currently, it provides 20 percent of the electricity in New York and owns and operates one third of the state's high-voltage transmission system.

At its core, the NYPA narrative tells us how much can be achieved when new thinking and enlightened public policy are combined with technological ingenuity and incredible determination. For that reason I am grateful that Professor Brynner set out to tell this valuable story. Looking ahead, we can use the lessons from this book to help us chart the future development of the power industry.

—Andrew M. Cuomo,
Governor, New York State

June 2016

Preface

I began researching the New York Power Authority (NYPA) in 2012, intrigued by an energy producer with its own brand of pragmatic idealism. The corporate histories that I read presumed considerable prior knowledge and offered no sense of the context in which NYPA arose and operates today. When it became clear that the research and interviews required to write such a book would only be possible from within the organization, I applied for a job opening in Corporate Communications, where, on my first day, I submitted a proposal to write this book. I correctly guessed that approval would not come quickly, which was just as well, since I had a great deal to learn about the Power Authority's origins and operations.

Since clever advertising can make almost anything look good, one is naturally skeptical that any organization can live up to its public proclamations. But the Power Authority doesn't advertise, which is a key reason why the vast majority of New Yorkers—my unscientific guess is 90 percent—do not know that the Power Authority exists or what it does. Although the utility fosters a robust relationship with communities neighboring its hydropower plants and smaller facilities, in most of New York City and beyond, NYPA might as well be a secret, despite its considerable efforts to communicate proactively and live up to its policy of transparency. It is largely thanks to that policy that my proposal for this book was approved.

I set out to examine the emergence of New York's public power in the context of the fierce and often unprincipled scrum among the private utilities, which had far-reaching consequences for the people and businesses of the state. I also wanted to understand what had postponed the development of the public utility for decades, and what had finally made it possible. But the most salient issue, given the destruction wrought by recent superstorms, is the urgent need for the world's power producers to arrive at sustainable practices and to pioneer new ones by exploring all the opportunities for renewable energy. The importance of clean energy at this late date in climatic degradation cannot be exaggerated. The Power Authority's leadership is rooted in its clean hydropower, though NYPA has been tasked over the decades to operate a handful of oil- and gas-burning plants, as well as nuclear.

For some readers, the birth of electrification may evoke the revolution in information technology, and the early dot-com era in particular. Surveying the endless litigation over patents and infringements, the parallels are often striking. But the greatest similarity is between two new technologies that rose from luxuries to necessities in just a few short years.

Jill Jonnes's *Empires of Light: Edison, Tesla, Westinghouse, and the Race to Electrify the World* (2003) provided an invaluable guideline to the introductory chapter here, and I am indebted to Ms. Jonnes for her excellent research. Some of the other works that contributed significantly to my understanding of that early era include Paul Israel's *Edison: A Life of Invention* (1998), David E. Nye's *Electrifying America: Social Meanings of a New Technology* (1990), Marc J. Seifer's *Wizard: The Life and Times of Nikola Tesla, Biography of a Genius* (1996), and Thomas P. Hughes's *Networks of Power: Electrification in Western Society, 1880–1930* (1983).

I owe a debt of gratitude to many people at NYPA, starting with the former Director of Corporate Communications, Michael Saltzman, who understood why this book needed to be written, and Ethan Reigelhaupt, who, as Vice President for Corporate Communications, has been very helpful during the completion of this work. I am especially grateful to Gil Quiniones, the President and CEO of the Power Authority, for authorizing this history, along with Governor Andrew Cuomo. With the assistance of Jill Murman-Payne and Connie Cullen in particular, I learned the ropes and began to make sense of the organization. My research and writing for NYPA's annual sustainability reports were valuable paths to learn how environmental issues guide decisions in the Power Authority's vast operations. And very special thanks to Michaela Walsh, who brought me together with Cosimo Books.

This book is indebted to the writers of NYPA's previous corporate histories, including Edward Eckstein, Cliff Spieler, Ira Fine, and especially Stephen Shoenholz, who proofread the manuscript more than once. Without their fine, meticulously researched work, this book would have taken more years to write than I could have devoted to it.

CHAPTER ONE

THE BIRTH OF ELECTRIFICATION

In 1907, Governor Charles Evans Hughes declared that the undeveloped waterpower of New York "should be preserved and held for the benefit of the people and should not be surrendered to private interests." Toward that end, the Republican governor secured passage of legislation commissioning "plans for the progressive development of the water powers of the state for the public use and state ownership and control." That seed of an idea produced the New York Power Authority, originally known as the Power Authority of the State of New York (PASNY).

The Origins of Public Power

Twenty-five years earlier, the first home in the state had been successfully wired and illuminated electrically. It should come as no surprise that this was the private home of J. Pierpont Morgan and that his electrician was Thomas Edison. Just as it is today, electrification was capital intensive in 1882, and entailed enormous start-up costs, particularly in the early years as it moved out of the laboratories into the streets, factories, and homes. It must have seemed inevitable that only the wealthiest individuals and companies would enjoy the benefits of electricity.

But even in the 1880s there was an alternative. Two years before J. P. Morgan's house was illuminated, the city of Wabash, Indiana, proudly lit its courthouse square with four powerful arc lamps. To accomplish this, the modest town of 3,800 with nary a millionaire established the nation's first public power company. The following year, the town of Butler, Missouri, followed suit; its arc lamps, mounted atop the Bates County Courthouse, could be seen twenty miles away. Privately owned utilities were spreading across the continent, selectively targeting cities with the greatest potential for profit. For other cities, municipal ownership began as a practical solution, mostly in small communities that were not profit-worthy. By 1890 there were some 150 municipally owned electric companies in the United States, and a decade later there were 800, amidst 2,200 private utilities.

There were other important advantages to public power. To begin with, it was significantly less expensive, since no profit was extracted from the revenue. And when a private power company was the only game in town, it could set a very high price to deliver electricity, often remaining the regional monopoly for years. By 1887 the operating costs of a private utility were double those of nonprofit local ownership, and those lower prices set by municipalities also provided a yardstick by which to measure the excessive rates of the private utilities. Local ownership also meant that residents' electricity payments would remain within the city or county, which helped the local economy.

Another tangible benefit of public power was more difficult to measure: all the decisions were made locally by elected officials under the close scrutiny of the town folk: siting the power plant for the convenience of the majority, downstream or downwind from the town; safely transporting and locating the flammable stores of fuel; devising a sensible system of transmission wires that avoided the threat of fire or electrocution. The impact of these decisions and dozens more upon the community could best be judged by ... the community. Environmental considerations were a factor with public power from its beginning, the same considerations that today are addressed by *sustainable practices* that meet economic, environmental, and social needs in ways that do not harm future generations.

State-Owned vs. Privately Owned Utilities

The fact that a staunch Republican governor like Hughes believed the state's water power "should not be surrendered to private interests," but instead should be reserved for "public use and state ownership and control," is a clear indication of his pragmatism. But it is also a measure of just how egregious the corporate free-for-all had become in the drive for electrical dominance, to the detriment of New Yorkers. Hughes clearly believed that low-cost electricity provided by the state of New York was a necessity, even though no other state was producing power yet. That same year Hughes created the first state regulatory commission for power, also bucking his party's enduring anti-regulatory policies.

Hughes recognized the need for a state-owned power company after having observed the birth of commercial electrification while studying at Columbia Law School. He was working in Lower Manhattan during the War of Currents, in which Thomas Edison laid siege publicly to the power-generating partnership of George Westinghouse and Nikola Tesla, who nonetheless prevailed, only to see their trophy seized by someone else altogether. This determinative moment in the evolution of power production was played out in New York State, giving Hughes and other New Yorkers a clear understanding that firm restraint and guidance over privately owned utilities was necessary for the public good and provided an alternative that the state could rely upon. The outcome of that "war" and other corporate shenanigans had convinced Hughes that unless the state created its own generating potential, it could not challenge the exorbitant fees exacted by private monopolies that can stagnate economic growth. Other departments of the state—water, waste management, emergency services—were crucial services for survival, but energy was the vital engine for growth, productivity, and prosperity.

Electrical wizardry had evolved through decades of international scientific collaboration. Now its ownership was parceled out as private property by overlapping patents and legalistic maneuvering, with the public interest a distant consideration at best. It was precisely during these twenty-five years, 1882–1907, that electricity evolved from a luxury for the wealthy to an urban and then a universal necessity. It was therefore, Hughes recognized, a responsibility of the state. In reassessing that era, one can see that the need for a state-owned utility was undeniable, made even more evident in light of the fierce opposition that delayed its realization for decades.

Harnessing and Understanding the Power of Electricity

Human ingenuity tamed and harnessed the elemental power of electricity in the late nineteenth century. The primordial force of lightning that had excited religious dread long before the age of Zeus soon became the lifeblood of mankind's endeavors, transforming daily life. Fifty years after the first commercial power, half the homes in the United States were wired. Today many of the most isolated communities in the world have small generators, even if they lack access to the large-scale "ethereal fluid" that powers factories, hospitals, subways, banks, supermarkets, and homes. The scientists and entrepreneurs who advanced electrification could not have imagined the many inventions it would unleash, without which the twenty-first century would be very much like the nineteenth.

One cannot reproach those innovators for failing to anticipate the cumulative impact of electrification upon our ecosphere. They could not have guessed that the human population would grow sixfold in the following hundred years, thanks in large part to electricity; but if they had, they probably would have supposed that by now we would be generating all our electricity by sustainable methods. Indeed, the first hydropower plant built at Niagara in 1895 demonstrated the way forward to clean generation and long-distance transmission of electricity.

Although that primordial power is now a commonplace utility, it remains a complete mystery to most of us. In the industrialized nations, as if by magic, we have unlimited power at our command, lurking in the wall behind the socket. Where that invisible force comes from rarely crosses our minds except when we are paying the monthly bill or there is a power failure. People don't generally pause to think about belching smokestacks or nuclear cooling towers while recharging their smartphones.

Because so few understand electricity, the colossal technology required to produce and deliver it reliably to all those sockets remains a kind of industrial magic in the minds of many. The genius of hydropower could well be mistaken for alchemy as it transforms a rushing current into an electric current. The wonderment of a child opening a refrigerator door to see the light inside switch on is not shared by most adults, but ponder this fact: power has to be generated *in the very same millisecond* that it is used. So, if the electricity comes from hydropower, when the child opens the fridge, at that instant a mass of water the size of a Buick (a *very* rough estimate) has to fall through a turbine to turn on the tiny light.

With so many millions of children and refrigerators, one can imagine what a challenge it is to "balance the load," producing just enough power to meet consumers' needs by continually fine-tuning the supply of electricity and anticipating the demand. A graphic illustration of this arose in England in the 1960s, where every Wednesday night there was a mysterious spike in demand just after 10 p.m. nationwide. The cause was soon determined: the popular television show *Coronation Street* ended at that hour, whereupon millions of British families plugged in their electric kettles for a cup of tea. Trouble was averted by boosting power generation at that moment each Wednesday to balance the load. Usually, determining the cause of a surge in the grid is much more complicated.

One reason for the complexity of power generation and distribution today is that it is not the result of a single, overarching plan for any state or region. Instead, electrification evolved by patchwork through corporate competition, advancing technologies, shifting demand, changing social environments, and progressive safety regulations. By a gradual accretion of discoveries and inventions, the modern system came into being, with new uses for electricity driving the growth in demand.

The Scientists Behind Electric Discovery

Man-made electricity was the collective achievement of scientists sharing their knowledge around the world for decades, but its whole history can fit in a few pages. In 1600 William Gilbert coined the word "electric" from "elektron," the Greek word for *amber*, which had been known since ancient Egypt to produce static electricity. He demonstrated that other substances, including glass, could also produce a magnet-like power of attraction when rubbed vigorously. In 1727 another Englishman, Stephen Gray, succeeded in aligning the chaotic energy of static into a current that flowed along 765 feet of wire.

Next, a scientist in Holland (and another in Germany) devised the Leyden jar, a primitive means of storing the electric charge generated by static. In 1752, Benjamin Franklin famously electrified a Leyden jar by flying a kite in a thunderstorm, proving that lightning and static share the same "electrical virtue." To store more static electricity, Franklin linked four Leyden jars together in what he named a "battery" because the shape of his device evoked a battery of cannon.

Electricity was only good for parlor tricks until 1800, when Alessandro Volta's superior battery provided a continuous current. Using voltaic piles, engineers experimented for years with different forms of telegraphy that required minimal electrical power to cover short distances. The collaborative scientific spirit took a turn toward the litigious as soon as the potential profits of long-distance communication became apparent, especially from government contracts. By 1838, Samuel Morse's system had prevailed, thanks largely to the relays he added to extend the reach of transmissions, as well as the efficiency of the binary Morse code he developed.

But lighting was the Holy Grail for electrification. In 1808, at the Royal Society, Sir Humphry Davy demonstrated the first "arc light," as he named it. Using a powerful voltaic pile he had built, Davy ran the current between two carbon rods, producing a bright bluish-white arc of light as the surrounding gases ignited, growing more intense as the rods were moved to four inches apart, and continuing to burn until the rods were consumed.

This was breathtaking to his first audiences who, in the age of whale oil and coal-gas lamps, had never seen anything on Earth so bright. But the carbon rods burned away in just two hours, and the Voltaic pile was also too short-lived for this first arc light to serve any practical purpose. Until some method was found for generating a steady source of electricity, the development of lighting had to wait. Davy's ongoing experiments with electricity were substantial, but perhaps his greatest contribution to science was his choice of a remarkable assistant named Michael Faraday.

Faraday was a self-educated young man with an overwhelming curiosity that later led him along many scientific paths, including the invention of benzene and the liquefaction of gases. But he was mostly drawn to Davy's electrical experiments. Faraday was also struck by the work of Danish physicist and chemist Hans Christian Ørsted, who for the first time had shown a relationship between electricity and magnetism after noticing that a compass needle reacted to a charged wire nearby. He went on to demonstrate how to create magnets with electricity. This discovery, along with the work of André-Marie Ampère in France, offered Faraday a promising avenue of research. He tried to reverse Ørsted's discovery without success until he discovered that moving a magnet back and forth within a loose coil of wire produces a brief charge, because it is the *motion* of the magnet, not the magnet alone, that generates electricity. He realized, in fact, that the magnet converts mechanical energy into electrical energy.

Faraday exhibited the world's first generator at the Royal Society in 1831. By spinning a disk made of copper (the best conductor, he found) between the poles of a horseshoe magnet, he could draw a steady current—drect current, or DC—from the disk and transmit it along a wire. This generator, or dynamo, was only a model for one that would have a practical use, which took another forty years to develop.

In the early 1870s, Belgian engineer Zénobe-Théophile Gramme, working in Paris, produced a much-improved generator that replaced natural magnets with electromagnets, as suggested by Werner von Siemens in Germany, and the result was a much stronger current. This was the basic generator that launched the revolution in lighting over the next decade. Gramme's second contribution was almost as significant. He showed that, while the movement of a spinning disk could generate electricity, the process was reversible, so that electricity could make the disk spin, producing mechanical energy that could be put to other uses. Gramme had thus invented the first electric motor.

But it was Gramme's generator that traveled swiftly across Europe and the Atlantic. At the Philadelphia Centennial Exposition in 1876, partners William Wallace and Moses Farmer presented their improved variation—the first American-made dynamo—and with this 8-horsepower, steam-driven generator, they demonstrated three versions of arc lights. Few of the Exposition's visitors seemed to remember them even a few years later, perhaps because their generator was dwarfed by the 45-foot-tall Corliss Centennial Engine that produced 1,400 horsepower and powered almost all the exhibits. One of those exhibits was the new multiplex telegraph system shown by a successful 29-year-old named Thomas Edison, who would become world-renowned a year later for his phonograph. But for whatever reason, the Wallace-Farmer exhibit attracted very little attention, and so a prime opportunity to promote electric lighting was lost. In fact, every World Fair and Exposition for the next fifty years featured the latest electrical innovations.

In Paris, Pavel Jablochkoff, a Russian engineer trained in St. Petersburg, devised the most successful arc light yet. In the "Jablochkoff Candle," when the first pair of carbon rods burned down to nubs, they ignited another pair, and then another, inside a globe of enameled glass. Collaborating with Jablochkoff, Gramme discovered that the pairs of rods burned more evenly if his dynamo delivered pulses of electricity instead of a continual flow. This single-phase "alternating current" was Gramme's third crucial discovery. The full significance of AC power would not be appreciated for years, but with it the Jablochkoff Candle could burn continuously for sixteen hours.

The Paris Exposition of 1878 represented the Promethean moment for arc lighting in Europe. Jablochkoff and Gramme were commissioned by the city to illuminate a half mile along Avenue de l'Opéra with sixty-four Jablochkoff Candles. This was the birth of electric lighting, an awe-inspiring event for the thirteen million visitors who experienced it. As one visitor wrote, "The whole street, to the tops of the loftiest houses, is ablaze with a flood of beaming light which makes the streets seem like the scenes of some grand play at the opera." In just a few years, several Paris neighborhoods were illuminated by Jablochkoff Candles mounted on poles high above the streets to reduce their powerful glare.

News spread swiftly that streets could be lit at night by white-hot electric light instead of the dim trembling yellow of coal gas, and inventors across the United States scrambled to join the fray. Charles Brush of Cleveland soon became the American leader in the field, manufacturing his own arc lights and a dynamo—a much improved version of Gramme's—that could ignite up to forty lights in one circuit.

By April 1881, according to *Scientific American*, the Brush Company had installed more than 6,000 arc lights around the country, and had contracts for two dozen more cities. The lights were best suited for outdoor use because their intensity would be unbearable in an average room and, as the carbon rods burned, they produced an alarming odor and a loud hiss. But in the absence of any alternative, by 1880 Brush and other companies were installing arc lights indoors at steel mills, textile factories, railroad depots, churches, lighthouses, and a large Boston clothing store. With improvements, Brush's carbon rods lasted 100 hours without replacement. In New York, he illuminated Broadway from Union Square to Madison Square, making it "the Great White Way" before that sobriquet worked its way uptown to 34th Street and later to Times Square and beyond.

Brush's first major competitor was the Wallace-Farmer Company. In the two years since the Philadelphia Exposition they had developed a different variation of the Gramme dynamo at William Wallace's foundry in Ansonia, Connecticut. There, during a visit, Thomas Edison decided he would invent a practical light bulb.

Edison and the Light Bulb

Jill Jonnes's *Empires of Light: Edison, Tesla, Westinghouse, and the Race to Electrify the World* (2003) offers the full, rich drama of the War of Currents in exquisite detail, along with perceptive biographical portraits of its protagonists and its unexpected winner. This corporate war of attrition and its aftermath made readily apparent the need for public power and the regulation of essential utilities, because the biggest loser of this conflict was the citizenry of New York. It also quickly became obvious that unbridled competitive capitalism was detrimental to any economic development, whether in a county, a state, or a nation. The indispensible need for public power arose directly out of the War of Currents as it played out in New York.

Edison, it is clear throughout the literature, embodies a character of many contradictions: brilliant, stubborn, relentlessly curious, driven. He could be farsighted and shortsighted at the very same moment; a man of meticulous science and reckless finances; noted for great warmth and cold-blooded calculation. Edison, the formidable inventor and engineer, was unashamedly in the service of Edison, the entrepreneur. "Anything that won't sell I don't want to invent," he once said. "Its sale is proof of utility and utility is success." That commercial pragmatism made his big dream possible.

Edison was 31 years old when he visited Wallace's foundry in September 1878, with a newspaper reporter close at hand. After spending his teenage years traveling the United States as a "tramp" telegraph operator, he devised a substantial improvement with multiplex telegraphy that he sold to the giant Western Union, owned mostly by Cornelius Vanderbilt. He had then patented the phonograph in 1877, which took Washington, D.C., and the country by storm. Now, as he studied the latest Wallace-Farmer generator and arc lamps, his imagination took fire and his vision was born. Before he left that evening, Edison bluntly told Wallace, "I believe I can beat you making the electric light. I do not think you are working in the right direction." He later wrote: "I saw that what had been done had never been made practically useful. The intense light had not been subdivided so that it could be brought into private houses." He immediately devoted his energies to creating a practical light bulb, along with all the publicity he could muster.

This talent for public relations served Edison's considerable ego, but more importantly helped him meet his greatest challenge: raising enough capital for his big dream. Even a perfect indoor light bulb was useless in houses that lacked electricity. His competitors were already installing dynamos that served individual buildings, but Edison began building a central power station that incorporated six locomotive engines and distributed electricity throughout whole neighborhoods. "He picked an area in downtown New York," wrote utilities expert Leonard Hyman, "because it had 1,500 gas light customers and the potential for 750 electric motors, so power could be sold both day and night. He wanted to serve the financial district for the prestige involved in doing so." The plant was located on Pearl Street, within sight of the nearly completed Brooklyn Bridge. It generated enough power for 1,200 sixteen-candle lamps and consumed *five tons* of coal and 11,500 gallons of water every day. Everything about it was expensive, including the eight miles of copper wire, which provided less electrical resistance than other metals but cost much more. The wire just for his first half-mile display cost $18,000. The central station alone cost three times the original estimate that he had quoted to his investors, J. P. Morgan, Jay Gould, Henry Villard, William Vanderbilt, *et al.*

So, well before Edison's light bulb had been perfected as a viable product, he invited his impatient Wall Street investors and the press to his vast private laboratory in Menlo Park, New Jersey, to see the power of his new generator, the "long-legged Mary Ann," and to raise the next round of cash he would need to wire part of Lower Manhattan. This second private exhibition and others that followed were smashing successes, as recorded by breathless reporters always glad to attend these junkets. His success in creating the first reliable incandescent bulb was trumpeted prematurely by headlines in the *New York Herald*: "The Great Inventor's Triumph in Electric Engineering." According to the article, "Edison's electric light ... is produced from a tiny strip of paper [through which] is passed an electric current, and the result is a bright beautiful light, like the mellow sunset of an Italian autumn.... Edison makes the little piece of paper more infusible than platinum, more durable than granite.... This light, the inventor claims, can be produced cheaper than that from the cheapest oil."

But because the glowing paper filament wasn't durable enough to succeed commercially, several of the lab's sixty employees continued testing other materials for months. Achieving a perfect vacuum in each bulb was another technological challenge that obliged Edison to first improve the design of the necessary pump. It is no wonder that Edison was consistently late and over-budget in delivering the innovations he promised his tycoon investors, because his ambitions never stopped growing. He was relying upon economy of scale: a larger electrical system would produce not only a greater return, but also a greater percentage return for each dollar invested.

Despite Edison's best arguments, "electricity was too expensive for the average customer and remained so until the early 1900s," wrote Hyman, and Wall Street became more wary as actual costs raced past all estimates, leaving just a handful of very wealthy, shrewd investors. Edison's biographer Paul Israel wrote that when he turned to the development of a commercial system, he "depended on the support of corporate capital. While Edison the individual is celebrated as the inventor of the electric light, it is the less visible corporate organization of the laboratory and business enterprise that enabled him to succeed." As Jonnes explained, "Edison's unique access to big corporate money—first through contracts with Western Union and then through Wall Street light bulb money—gave him an enormous advantage over his rivals.... Edison was inventing not just the light bulb, but a new kind of relationship—however prickly and difficult—between corporate capital and scientific creativity." In the end, it was corporate capital that swept him away from the very field he pioneered.

There didn't seem to be any alternative to Wall Street, given the limitless capital he could imagine spending. Public power for indoor lighting was not even considered, and the state government in Albany had no reason to be involved. As for New York City, although Boss Tweed himself had recently died in jail, the endemic corruption of Tammany Hall Democrats was not what Edison wanted to partner with.

He did, however, woo the city's aldermen and officials to a banquet at Menlo Park to see his latest incandescent lighting. He knew he would need their cooperation, and the mayor's, not only to obtain a franchise to install his system, but also to bury his wiring under the city streets. Wherever possible, he planned to follow the same channels already dug by the gas companies—the very utility that his light bulb was designed to eliminate. He did not want his wires to join the web of telephone wires draped above the streets. New York already had, in the words of historian David Nye, "a vast, intricate network of wires over all the city especially in the business sections; and the avenues and streets showed a forest of tall poles, many of them carrying several hundred wires.... As these poles necessarily differ in height, the wires upon them form a complete network, rendering the efficient use of the hooks and ladders and life-saving apparatus of the fire department almost impossible." Edison was rightly concerned that his system could be compromised by such jerry-rigging and by the threat of electrocuting passersby. In fact, one of his strongest selling points was the safety of his electric lighting compared to gas jets, which routinely caused accidental gassings and terrible fires.

By this time, the *New York Evening Post* reported, there were already "six different companies at work introducing electric lights in this city, the lights being known as the Brush, Maxim, Edison, Jablochkoff, Sawyer, and Fuller (Gramme patents) lights." Maxim, who had unabashedly copied Edison's earliest patented bulb, probably expected an infringement suit, but his illumination was installed two years ahead of Edison's. Most of them had devised their own generators, at different voltages, some of which worked better with motors than lighting. New York City's electrification might have evolved like London's by 1900, writes historian Thomas P. Hughes, where sixty-five separate utility companies were operating with small, coal-fired plants that generated "ten different frequencies, thirty-two voltage levels for transmission and twenty-four for distribution and about seventy different methods of charging and pricing."

By June 1882, when Edison installed a private generator and illuminated J. P. Morgan's home, his shop had redesigned every element of electrical generation and transmission, reducing to one-eighth the amount of copper wire he needed through a feeder network of much thinner wire. And through trial and error, testing hundreds of materials, including bamboo, the most successful filament proved to be a simple cotton thread coated with lamp carbon. Thanks largely to patent attorney Lewis Latimer, Edison was able to defend his patent.

And on September 4, 1882, the entire system was ready. Edison himself was ready to pull the switch, though not actually at the Pearl Street plant. Tellingly, he was standing beside J. Pierpont Morgan at the tycoon's Wall Street office, where, at exactly 3:00 p.m., he threw a symbolic switch while simultaneously an employee at his central station closed the actual circuit breaker. Across the wealthiest swath of Lower Manhattan, 400 lights went on, serving eighty-five customers, thanks to the first electricity from an investor-owned utility.

The total investment, provided largely by Morgan and railroad mogul Henry Villard, had risen to $500,000. Hyman wrote that Edison's plans were "overambitious, schedules fell by the wayside, the project came in at a cost several times over the estimate, and meters were not ready, all of which resulted in no bills for months." Morgan also saw clearly that selling individual generating units like the one in his own home would produce more profits for Edison Lighting than these enormous central plants, with the added benefit that electricity would remain a privilege for the well-to-do, as he believed it should. Nevertheless, in spite of the problems and disagreements, the commercial distribution of electricity was now a going concern.

CHAPTER TWO

THE WAR OF CURRENTS

The giants of the electronic revolution might have been ready to distribute electricity to homes and businesses far and wide across the United States, but their efforts weren't going to go very far. Not with direct current.

AC vs. DC

As early as 1878, when Edison had hastily designed and then slowly refined the light bulb, his plan was dictated by the price of copper, with which the history of electrification has always been inextricably entwined. Edison was committed to his 100-volt "pressure," which could be transmitted over much thinner copper wire; even his all-important discovery of cotton fiber as filament was predicated upon the cost of copper wire for transmission. He therefore designed the generators and every patented device in the system for 100 volts of direct current, knowing full well that this low-voltage direct current lost so much power from resistance as it traveled that it could never go much beyond a half mile from the central station. To provide the whole city with power, he would have to build *dozens* of central stations, one every ten blocks or so, each with six locomotive engines and requiring thirty tons of coal and 11,500 gallons of fresh water *every day*. And of course every station would produce a pall of black, particulate smoke and require an endless convoy of horse-drawn coal wagons through the snarled, boggy, trash-ridden streets of Manhattan.

This was same year Jablochkoff and Gramme demonstrated, to great acclaim, an early version of their "alternating current" with the arc lights at the Paris Exposition. By 1882, when Edison turned on the lights in Lower Manhattan, many advances had been made in AC generation, and its ability to be transmitted much further had been shown, especially in London and Budapest.

It was in Budapest in 1882 that a young Serbian engineer named Nikola Tesla had a vision as he strolled beside the Danube River with a friend. "I drew with a stick in the sand," he later wrote. "The images I saw were wonderfully sharp and clear." He had perfectly imagined a motor that would run on alternating current. He turned to his friend and said, "Isn't it beautiful? Isn't it sublime? Isn't it simple? I have solved the problem. Now I can die happy. But I must live, I must return to work and build the motor so I can give it to the world. No more will men be slaves to hard tasks. My motor will set them free, it will do the work of the world." As a journalist who interviewed him later wrote, "He believes that electricity will solve the labor problem.... It is certain, according to Mr. Tesla's theories, that the hard work of the future will be the pressing of electric buttons." Tesla's prediction proved stunningly accurate for much of the workforce.

What Tesla envisioned beside the Danube was a motor which, when running in reverse, would generate a polyphase alternating current. That is, wrote his friend and biographer John J. O'Neill, instead of using a single AC circuit, "what Tesla did was to use two circuits, each one carrying the same frequency, but in which the current waves were out of step with each other, hence polyphase. This was the equivalent of adding a second cylinder to an engine." Jonnes explains that "Tesla planned to so position his currents that, as the first waned, the next would kick in, creating an invisible whirling magnetic field ... that some would later call a 'wheel of electricity,' with almost no wearing parts. However, all this was still firmly in Tesla's head."

It would be more than a decade before Tesla could project his vision into the real world, remarkably enough at Niagara Falls, and on a much grander scale than he could have hoped for. Along his solitary, extraordinary life, time and again this sensitive, naïve genius lost everything he owned: to a thief the day before he boarded his ship to New York; again when his workshop burned down years later; to the financial manipulations of almost all his investors; and again in his final years, when he relied upon the charity of others.

Despite many early misfortunes, by 1884 Tesla was working in Thomas Edison's shop in New York, after being recommended by Edison Lighting in Paris. "He and Edison were like oil and water," wrote Jonnes. Tesla later critiqued his employer's process of elimination as "empirical dragnets": "If Edison had a needle to find in a haystack, he would proceed with the diligence of the bee to examine straw after straw.... His method was inefficient in the extreme ... and, at first, I was almost a sorry witness of such doings, knowing that a little theory and calculation would have saved him 90 percent of his labor." Relations between the two men soured quickly after Tesla completed a major undertaking for which Edison's manager had promised him $50,000. According to Tesla, when he asked for the money, Edison replied, "Tesla, you don't understand our American humor," and insisted that the offer had only been a joke. That scotched all trust between the two. Edison, who had been impressed with Tesla's hands-on skills more than his high-flown visions, insisted that "Tesla is the poet of science" whose ideas were "magnificent but utterly impractical." Tesla said that when he explained his plans for alternating current that could travel almost any distance, Edison answered bluntly that "he was not interested in alternating current; there was no future to it and anyone who dabbled in that field was wasting his time; and besides, it was a deadly current whereas direct current was safe."

It is not surprising that Edison was so committed, because his direct current system fulfilled its basic promise. Perhaps more importantly, his patents were all for DC equipment, for which others would pay royalty in future systems. Two years after Pearl Street began generating, his company had sold 378 smaller, isolated power plants, similar to the one in J. P. Morgan's mansion, in towns across the country, especially those that lacked gas lighting or did not appear profitable to the private utilities. Banks were not yet willing to finance electrical generation, perhaps waiting to see if this was just a passing fad. So Edison's company often loaned the money to its small-town customers to complete the sale; but his own investors balked at this. As for the large, central stations, only eighteen were sold by the end of 1884, and before long the inventor considered the finances of Edison Electric Light Company "a leaden collar." The smaller generators were more profitable than the central plants, so Morgan pressed Edison to concentrate on selling these. Still, "the Edison system," wrote historian Hughes, "had a major technical flaw—the extreme expense of distributing electricity at low voltage." Choosing to stay with low-voltage direct current because of the cost of copper wire was proving to be a false economy.

For the next few years Edison Electric Illuminating Company (Edison's company arm responsible for the creation of central stations) continued to add prestigious individual organizations to its roster of customers, including the New York Stock Exchange. But few cities had populations dense enough to attract hundreds of subscribers within half a mile of a central station, and with fewer customers, each one had to pay more, further limiting the service to the most affluent.

Still, Edison remained obdurate in his opposition to alternating current, even when its advantages were being proven around the country. In Massachusetts, a former shoe salesman named Charles Coffin led a group of investors who owned the Thomson-Houston system of arc lights (which freely "borrowed" from Edison's plans), adding to a daisy-chain of patent infringement suits and countersuits. All the players in the high-stakes world of power bought, sold, swapped, copied, "improved," hoarded, and defended patents almost on a daily basis, and Edison deeply resented the "patent pirates" who did, especially Thomson-Houston—in the person of Charles Coffin—and the latest newcomer in the field of electrification,

George Westinghouse, a young industrialist from Pittsburgh who premiered his DC system for lighting and motors at the Philadelphia Electric Exhibit in 1884.

George Westinghouse had made a considerable fortune in his twenties with numerous improvements for the railroads, most importantly his air-brake system for controlling each car of a train simultaneously, which before long became the standard for all railroads. From his work improving upon the oil lamps used for railroad signals, he focused all his efforts upon electric lighting. Joining up with engineer William Stanley, who had patented a silk-filament light bulb, he began buying up patents from Swan Incandescent in England and other companies. But after two years he turned to alternating current, and bought the rights to AC generators designed by Gaulard and Gibbs in England, Ganz in Hungary, and Siemens in Germany.

Westinghouse himself, together with his assistant engineer, devised the key element for his new system: a transformer that could increase or lower the voltage of AC current, which cannot be done with DC. With less resistance at the higher voltage, power could travel much further; then, with a transformer, it could be stepped down to the lower voltage required for light bulbs or small machines.

In 1886, Westinghouse's partner Stanley tested their first generator in Great Barrington, Massachusetts. Driven by a 25-horsepower steam engine, it transmitted a current of 500 volts that was stepped down to 100 volts by a transformer in each subscriber's building. This AC central station, the first built in America, could transmit high-voltage power over extensive distances. Within a month they had dozens of customers in Great Barrington; an older Edison station down the street had only six.

This made Westinghouse the leading champion of alternating current and *de facto* foe of Thomas Edison, who had the distinct advantage of worldwide renown. But high-voltage AC transmission had greater advantages. To begin with, customers did not have to live within eight blocks of a noisy, smoky power plant to have electricity. Better still, from the start the plant could be planned and built in an optimal location for coal deliveries, by rail or barge. But ultimately the greatest advantage was the lower price that Westinghouse could charge, thanks to a larger number of customers, owing to the greater range of AC. The limitations of DC transmission enforced the exclusive nature of its service.

Along with private utilities, the number of municipal power companies was rising by the year, especially in towns from the Great Lakes down through the Great Plains. In each case, elected town folk with little or no scientific understanding, who had never seen electric lighting unless they'd been to a World Fair, had to make the difficult and costly choice between AC and DC. As salesmen from all the companies fanned out across the nation, the advantages of AC were obvious to each new town that was approached: AC would obviously benefit more of their town's citizens, and if they could establish nonprofit local ownership, the operating cost would be half that of private utilities, according to an 1887 survey. Edison had already built 121 central stations around the country, but in just his first year, Westinghouse had sixty-eight central stations completed or under contract, and Charles Coffin (the Thomson-Houston Co.) had another twenty-two contracts for their AC system, which incorporated transformers they licensed from Westinghouse.

But Edison's name was golden; better still, it was illuminated. Everyone had seen photos of his name spelled out in bulbs at the Paris Exhibition in 1881, the first time that had ever been done, launching a deluge of lighted advertising that soon outshined the street lights in big cities, as it often does today.

As the future prospects of electrification became clearer, the price of copper reached a peak in 1888, and then collapsed after a failed attempt by a Frenchman to corner the global market, soon followed by a similar attempt by the Rothschilds in England and France to take over the Anaconda mines in Montana, followed by the Rockefellers' success when their Amalgamated Copper Mining took control of Anaconda. This alternate path to a controlling monopoly over electrical generation would later attract antitrust regulation under President Theodore Roosevelt.

Westinghouse and the Electric Chair

Edison could do nothing to extend the range of his central stations beyond a mile, at most. So instead, to protect his enormous investment in DC, he began to declare publicly what he sincerely believed in private at the time. "Just as certain as death," he wrote to an associate, "Westinghouse will kill a customer within six months after he puts in a [AC] system of any size.... It will never be free from danger." Edison's low-voltage transmission lines posed no such threat, and it seems he was firmly convinced that poorly designed AC systems would retard electrification as a whole.

Edison now devoted his energy and his fame to besmirching the reputation of alternating current, the Westinghouse Company, and its practices. A pamphlet entitled "A Warning from the Edison Electric Light Co." aimed 84 pages of vituperative squarely at any power company that threatened citizens with high-tension wires over the streets, cataloguing the handful of deaths that had occurred from fallen wires. And it contrasted Edison's direct current with that of the Westinghouse Indirect or Converter System. "Both systems start from the coal pile; both employ steam boilers and engines." But aside from its relative safety, the biggest advantage to direct current was the *motors* it could run, as well as the lighting; no one had yet produced an AC motor. And Edison, boasted the pamphlet, would soon also produce "heating by electricity ... the next immediate practical application of electricity on a vast scale." In actuality, the next practical application was something altogether different.

In late 1887 Edison received a providential letter from Dr. Alfred Southwick, who served on the New York Death Commission. He was seeking advice for an electrical method to provide a "less barbarous" form of execution than hanging. Gradually Edison was recruited into Southwick's widening crusade, and finally replied that "the most suitable apparatus for the purpose is that class of dynamo-electric machinery which employs intermittent currents. The most effective of these are known as 'alternating machines.' The passage of the current from these machines through the human body even by the slightest contacts, produces instantaneous death." Since the word "electrocuted" had not yet been coined, Edison suggested one could say that those executed had been "Westinghoused" or "condemned to the westinghouse."

Another ardent opponent of alternating current, Harold Brown, joined forces with Southwick and Edison, after writing an outraged letter to the *New York Times* declaring that "the 'alternating' current can be by no adjective less forcible than damnable.... The public must submit to constant danger from sudden death in order that a corporation may pay a little larger dividend." Brown began giving lectures during which, without prior warning, he electrocuted a variety of dogs at different voltages of AC, while members of his audience often fled in horror and disgust. That did not faze Brown, who went on to publicly electrocute calves and horses, all the while being secretly paid by Edison. Brown persuaded the state legislature to formally designate AC as New York's "executioner's current," highlighting its purported danger. Brown was named as the electrical expert for New York's prisons in what may have been the state government's first contract for electricity.

With Edison's surreptitious assistance, the first electric chair was built and a used Westinghouse generator was installed upstate at Auburn Prison, where one William Kemmler was to be executed for the murder of his wife. When his attorney appealed against this "cruel and unusual" manner of death, Edison himself testified in court that, using the Westinghouse dynamo, Kemmler's temperature would rise "and the heat would evaporate all the fluids in his body and leave him like a mummy." The next day the headlines blared: "Edison Says It Will Kill ... One Thousand Volts of an alternating current Would Be Sufficient." Edison must have counted this a great triumph: he had personally placed the danger of AC on the front page of *The New York Times*. After the grisly electrocution of a linesman working on AC arc lamp wires, Edison decreed flatly, "I can write upon this subject only as one convinced.... My personal desire would be to prohibit entirely the use of alternating currents. They are as unnecessary as they are dangerous."

On August 6, 1890, William Kemmler was led from his cell to the newly arranged death chamber in the basement of Auburn Prison, where twenty-four witnesses watched as he was strapped into the chair and had electrodes connected to his body. With little ceremony, the switch was thrown in the dynamo room and the current raced through Kemmler's stiffened body for seventeen seconds. But when they detached him, he was still breathing. They reattached him for several more minutes, effectively roasting the man. As several witnesses became sick from the smell and others fainted, Southwick from the New York Death Commission declared, "This is a grand thing, and is destined to become the system of legal death throughout the world." But a sheriff from Buffalo was so "badly affected" that he took to his room and, reported the *Times*, "bitterly denounced the electrical execution."

Despite public horror at reports of Kemmler's grotesque death—and some suspicion that Edison had devised this tortuous appliance to damage his business rivals—after serious tweaking, the electric chair did indeed become the principal means of execution in the United States for most of the twentieth century. Meanwhile, in news articles and corporate dispatches, Edison continued to throw the full heft of his prestige against his foes, disparaging AC for its supposed dangers, high cost, and inefficiency, along with its one undeniable drawback: the lack of a motor that could operate on alternating current.

The Science of Electricity

In 1888, six years after Nikola Tesla had envisioned an AC motor while strolling beside the Danube, he was able at last to demonstrate his newly patented machine to the public at Columbia College. Soon after, George Westinghouse invited him to Pittsburgh and agreed to pay Tesla $20,000, a generous sum, and generous royalties for the device. For the next year Tesla worked at the Pittsburgh factory, trying to coordinate his machine with the Westinghouse AC generators that operated at double the 60-hertz frequency of his own.

It is important to note that as Tesla and Westinghouse and Edison sought to harness this useful and lethal power, they had no idea what constituted electricity. The essence of this "ethereal fluid," "invisible agency," "subtle, vivifying fluid," this "electrical virtue" remained a mystery. An electrical current, *we* know, is a continuous flow of free electrons; but in the 1880s the electron was unknown and unnamed. The very definition of the word "atom" was "the basic unit of matter," so "subatomic particle" would have been an oxymoron. No one had determined the size of atoms, much less their composition of protons, neutrons, and electrons.

These three men were empirical innovators testing different means of producing desired phenomena, not theoreticians driven by curiosity about particle physics. Of course, they understood that the electromagnetism they were struggling to produce and control was an important natural phenomenon related somehow to both the Earth's polar magnetic force and to lightning. But they would have been flabbergasted to learn that electromagnetism is one of the four fundamental interactions (along with gravity, the "weak nuclear force," and the "strong nuclear force") that produced the universe.

In fact, the electromagnetism this trio of innovators created is the selfsame glue that binds electrons to nuclei and atoms into molecules: *it holds matter itself together*. By 1905, Einstein had deduced the Special Theory of Relativity based upon the "electrodynamics of moving bodies" that defined the underlying relationship between space, time, and the speed of light. Yet Einstein's theoretical work was only possible thanks to practical experimentation by the inventive minds that produced the dynamo, the current, and the light bulb. Beginning with Michael Faraday, it is the most wondrous achievement of collaborative science, which had the most far-reaching consequences, while creating a host of new responsibilities.

Although Tesla knew nothing of cosmology, he captured the enormity of his work's significance instinctively during a revolutionary lecture in which he anticipated Einstein by more than a decade. "Of all the forms of nature's immeasurable, all-pervading energy, which ever and ever change and move, like a soul animates an innate universe, electricity and magnetism are perhaps the most fascinating.... There must be a constant quantity of [electricity] in nature; that it can neither be produced nor destroyed....

The field is wide and completely unexplored, and at every step a new truth is gleaned, a novel fact observed.... We are whirling through endless space with an inconceivable speed, all around us everything is spinning, everything is moving, everything is energy. There must be some way of availing ourselves of this energy more directly."

Westinghouse and the AC Motor

Westinghouse was renowned for his integrity, determination, and persuasiveness. Biographer Henry Prout wrote that, "with his soft voice, his kind eyes, and his gentle smile, he could charm a bird out of a tree." His decency toward his employees was renowned; indeed, his was the first major company to allow a half-day off on Saturdays. Business partners also spoke of his congenial nature. But after extending a friendly opening to Edison with a personal note, which was flatly rebuffed, Westinghouse began replying with salvos of his own, publicizing the number of coal fires at Edison plants and other mishaps. He also countersued Edison for patent infringements with his light bulb, which he firmly believed was distinctly different, and he eagerly awaited a higher court ruling.

Antagonism did not come naturally to George Westinghouse, whose character Tesla described with great affection. "Though past forty then, he still had the enthusiasm of youth. Always smiling, affable and polite, he stood in marked contrast to the rough and ready men I met. Not one word which would have been objectionable, not a gesture which might have offended—one could imagine him as moving in the atmosphere of a court, so perfect was his bearing in manner and speech. And yet no fiercer adversary than Westinghouse could have been found when he was aroused. An athlete in ordinary life, he was transformed into a giant when confronted with difficulties which seemed unsurmountable." Still, litigious victory was not what Westinghouse went into business for. "A dollar given to a man does him ten dollars' worth of harm, while a dollar earned by his own efforts does him ten dollars' worth of good," he said in a rare interview, "so my ambition is to give as many persons as possible an opportunity to earn money by their own efforts," by building companies that are "large employers of labor." As his biographer Prout wrote, Westinghouse was always after progress above profit. This was perfectly in tune with Tesla's unselfish proclamation: "I must build the motor so I can give it to the world. No more will men be slaves to hard tasks. My motor will set them free, it will do the work of the world." Edison's signature maxim is a shade less altruistic: "Anything that won't sell I don't want to invent; its sale is proof of utility and utility is success." In short, Westinghouse wanted to give the world jobs, Tesla wanted to ease the world's toils, and Edison wanted to sell useful appliances.

But by 1888 even Edison must have understood that DC was only useful to the very wealthy; and with its limited range, DC could not benefit as much as AC from economy of scale, in which he placed such faith. Certainly his club of investors understood that AC could produce a greater percentage return for every dollar invested, but though they urged Edison to work with AC, he would not be budged. So Henry Villard, the Wall Street mogul who was Edison's leading investor and now also the President of Edison General Electric—a merger of all Edison's electric companies financed by Drexel, Morgan & Co.—began a back-room effort to force a merger with Thomson-Houston, led by the former shoe salesman Charles Coffin (who had repeatedly been rebuffed by Westinghouse). Apart from J. P. Morgan and his investor group, none of the parties wanted this merger. But Edison owned only 10 percent of his company. Charles Coffin finally demurred to Villard on the condition that he would be the leader of the merged giant of Thomson-Houston with Edison General Electric. Thomas Edison had been "Morganized." To add insult to injury, in a virtual dethroning, his name was stripped from that of the new company, General Electric.

Surely, wrote the editor of *Electrical Engineer*, wasn't the underlying cause of this devastating blow "the attitude taken, and persistently held, by Mr. Edison towards alternating current distribution? He could see no merit in that system. But upon its advent, its possibilities were promptly perceived by others.... Since its introduction for long-distance service, six years ago, it has practically driven the direct system from the field.... Mr. Edison ... has sought on every possible occasion to discredit it through the weight in the community of his justly great name. But the tide would not turn back at his frown."

Westinghouse, on the other hand, had no individual investors: he owned his company outright, though with support from various bank loans for specific operations and research. But in 1890 the financial world began a steep downward slide when the staid old Barings Bank in London almost imploded; as Argentina came close to defaulting on its national debt, it had exposed Barings's slim cash reserves. Credit froze up, and suddenly George Westinghouse was unable to meet his obligations. That also meant he could not afford to continue Tesla's research effort to mesh his AC motor with the Westinghouse AC generator. The bankers were spooked and insisted they would withdraw their support without a better potential payoff.

After a discussion with Westinghouse, Tesla decided to forgo his contractual royalty of $2.50 per watt sold, thus sweetening the investors' prospects and saving the entire project. With that, August Belmont, who represented the American interests of the Rothschild banking empire, provided the necessary funds. Almost two years had been wasted before Tesla returned to Pittsburgh to build the first AC motor, but the fundamental problem remained: his motor could only work at a lower frequency than the 133 hertz that the enormous Westinghouse generators were designed to produce. Finally, the Westinghouse men agreed to lower the frequency of the generators, as one of them later explained, "Strenuous efforts to adapt the Tesla motor to [the prevailing] circuit were in vain. The little motor insisted on getting what it wanted, and the mountain came to Mohammet."

The Bid for the World Fair

The final *champ de guerre* in the War of Currents was the Chicago Columbian Exposition celebrating the four hundredth anniversary of Christopher

Columbus's voyage. Like all the World Fairs, this would be the best possible showcase for electrical innovations, to establish the superiority of General Electric or Westinghouse. The two companies faced off in a bidding war for the job of illuminating the whole fair: installing 92,000 arc lights outdoors on almost 700 acres of swamp, which 7,000 workers were transforming into the fairgrounds. General Electric's preliminary bid was $1,720,000, but as soon as Westinghouse entered the fray, Charles Coffin dropped GE's price to one third that amount. Since he knew that several members of the city's selection committee were GE shareholders, Coffin was feeling confident until, at a public ceremony, the city opened the Westinghouse bid for the job: $499,550. The contract went to Westinghouse, after warnings from GE that he couldn't use their light bulbs.

Westinghouse set up a factory in Pittsburgh to produce 250,000 bulbs of his own design, expecting to lose General Electric's infringement suits against him. In December 1892, that is exactly what happened. Once and for all, the US Supreme Court ruled that Thomas Edison was the inventor of the filament bulb; General Electric had won the future profits, but Edison kept the considerable boasting rights.

And so, when the Chicago World's Columbian Exposition opened in May 1893, Westinghouse had built twelve generators, each weighing seventy-five tons, which themselves drew many of the fair's twenty-eight million visitors. Together they generated three times more electricity than the whole city of Chicago; to make this array more environmentally friendly, the numerous generators were fueled by oil instead of coal. Everything about the Exposition was gargantuan, starting with the 294-foot-tall wheel, the first built by Mr. Ferris to provide a bird's eye view. There were almost 60,000 exhibits, and at the center of Electricity Hall, two hundred yards long, stood a 45-foot-high. high monument to the company that was illuminating the whole Exposition, and was now the dominant generator and distributor of electricity: Westinghouse Electric & Manufacturing Co. Tesla Polyphase System.

Almost as a coda to the War of Currents, in the last days of the Exposition, Tesla gave a lecture that attracted a thousand electrical engineers, "the great majority of whom," according to the *Chicago Tribune*, "came with the expectation of seeing Tesla pass a [AC] current of 250,000 volts through his body" as the ultimate repudiation of Edison's campaign about its dangers. Tesla used the power running through his own body to power his new *wireless* bulbs. These bulbs represented the next step, going beyond Edison's bulb in lighting efficiency because they did not waste power generating heat; instead they were long cylinders without any filament, filled with fluorescent gas.

The Chicago Columbian Exposition of 1893 provided the full recognition rightfully owed to Thomas Edison for his invention of the light bulb, above all. But it was a hollow victory, since he had lost the War of Currents. Much worse, he had lost ownership of his own company to the autocratic monopoly of J. P. Morgan, who had given operational control of it to the "patent pirate" Charles Coffin of Thomson-Houston, whom Edison despised. Twenty years later, when he developed a new kind of battery, he declined to enlarge his operation by seeking investors on Wall Street. "My experience over there is as sad as Chopin's 'Funeral March,'" Edison acknowledged. "I keep away."

For the Westinghouse-Tesla generating system, the Colombian Exposition represented a victory lap. Their universal system, providing both AC and DC, offered an astonishing economy of scale that reduced costs at every level, and came to set the standard for the industry. But the crowning achievement of their alternating current would come three years later with the first long-distance transmission of power from Niagara Falls.

CHAPTER THREE

THE BIRTH OF HYDROPOWER

Converting Water to Electricity

Waterpower is as old as the pyramids, almost exactly. It also developed independently in ancient China, Mesopotamia, and Persia. By the 1700s, the energy of streams and rivers was used to mill grain to flour, saw logs into planks, and weave cloth from thread. In the 1820s, the city of Lowell, Massachusetts, was planned and built as a textile center between the Merrimac and Concord rivers, whose passing energy, collected by waterwheels, was transmitted to shafts spinning overhead in the workshops, from which belts carried the power to weaving and sewing machines below.

The feat of converting waterpower to electrical power must rightly be credited to William George Armstrong of Northumberland, England—the most environmentally forward-thinking engineer, scientist, and industrialist in the history of electrification. After studying a waterwheel, Armstrong was inspired to design hydraulic cranes that lifted boats onto dry dock using water pressure. Known as the "Magician of the North," he built an empire manufacturing these cranes, for which he was awarded the title of Baron. In 1863, he stood alone in foreseeing the drawbacks of coal, which, in his opinion, "was used wastefully and extravagantly in all its applications." Armstrong insisted, "England will cease to be a coal producing country within two hundred years," and he may yet prove right. He championed instead renewable energy in every form. Fully 150 years ago he calculated that "the solar heat operating on one acre in the tropics would ... exert the amazing power of 4,000 horses" and speculated that the "direct heating action of the sun's rays" could serve "in complete substitution for the steam engine." According to ecologist Polly Higgins, he believed that "the future lay in harnessing the forces of water, wind and sun, but recognized that the uptake of renewable power was dependent on the end of the use of fossil fuels."

This prescient gentleman was advocating hydroelectricity a decade before Edison used coal to power the first central station. In a high-profile London speech, Armstrong anticipated both the value of Niagara and the need for alternating current: "Whenever the time comes for harnessing the power of great waterfalls, the transmission of power by electricity will become a system of great importance." In 1878—four years before Edison used coal to light the bulbs in Morgan's mansion—Armstrong was practicing what he preached on his own 1,700-acre estate, damming a stream to create the world's first hydroelectric station, with which he powered an arc lamp, a roasting spit, a dumbwaiter, and even a proto-dishwasher in his home. Armstrong also planted seven million trees on his property.

In 1890, a large-scale Westinghouse-Tesla AC generator was inaugurated in Oregon, driven by a waterfall along the Willamette River that supplied power to Portland, thirteen miles away; but since Tesla's AC induction motor was not yet functional, the generator's application was mostly limited to lighting the city. The first test of the entire Westinghouse-Tesla polyphase system driven by hydropower came in 1891, in an exceptional situation that only alternating current could solve. Officials at the Gold King Mine in Telluride, Colorado (population 786) asked Westinghouse for an AC generator that could drive their stamping and crushing mill, but only if it could be powered by a waterfall three miles away. Indeed it could, producing the necessary 3,000 volts, which was stepped down to 100 volts to drive Tesla's new AC motor. The combined system handily proved its merit by operating without a problem for years, demonstrating that the Tesla's motor was ready for heavy-duty assignment.

Niagara

The mighty Niagara River drains the water from the upper Great Lakes: Superior, Michigan, Huron, and Erie. Between Lake Erie and Lake Ontario the river descends 326 feet, including the spectacular plunge of 182 feet at Niagara over the American Falls. Niagara Falls is the ultimate example of the *sublime*, which Immanuel Kant defined as a natural phenomenon that "does violence to the imagination." Clearly that describes the inspiration Charles Dickens felt when he visited the Falls in 1842. "It would be hard for a man to stand nearer to God than he does here," he wrote. "There was a bright rainbow at my feet; and from that I looked up to—great Heaven! To what a fall of bright green water! The broad, deep, mighty stream seems to die in the act of falling; and, from its unfathomable grave arises that tremendous ghost of spray and mist which is never laid, and has been haunting this place with the same dread solemnity—perhaps from the creation of the world."

The Niagara was used to drive a saw mill by Daniel Joncaire as early as 1757. When the Duke de la Rochefoucauld visited from Paris in 1799, he observed that a mile above the Falls "two saw mills have been constructed in the large basin, formed by the river on the left.... [The water] conveys the logs into the lower part of the mill, where, by the same machinery which moves the saws, the logs are lofted upon the jack and cut into boards." Since most communities evolved beside rivers, new settlements always constructed saw mills first, to cut boards for buildings.

There were early attempts to produce power from Niagara that bankrupted more than one family, but in 1879 Jacob Schoellkopf, who owned tanneries and flour mills, purchased land adjoining the Falls and commissioned Charles Brush to build a powerhouse. Brush installed a DC generator powered beside the river, which illumined his arc lamps and were trained on the American Falls, as well as sixteen street lamps in the slightly seedy tourist town of Niagara Falls. Schoellkopf's plant and its enhanced successor continued to operate for almost eighty years.

Once again the electrical destiny of a region was determined by J. P. Morgan. In 1889 the Niagara River Hydraulic Tunnel, Power, and Sewer Co., which had struggled for three years to raise capital, sought his investment. A Rochester engineer, Thomas Evershed, had proposed to divert water a mile above the Falls that could provide mechanical water power to as many as two hundred potential factories and mills, before returning to the river through a 7,500-foot, brick-lined tunnel to be dug directly beneath the town of Niagara Falls, population 500. All the necessary permits had already been obtained, along with a special state law authorizing the effort. But Morgan was only interested in the potential for electrical power. By his calculations, to be profitable the project would have to transmit electricity to the 250,000 residents of Buffalo twenty-six miles away. Although 200 hydroelectric plants were operating in the United States by 1889, nothing on this scale had been attempted. Morgan recommended that Edward Dean Adams be invited to head the project, to determine if such a venture was possible, technologically and financially.

Adams, descended from the presidents Adams, was himself a financier with a background in restructuring railroads. He had been an early partner of Morgan's investment in Edison Electric and helped him seize control of it. His mansion was conveniently located near Morgan's on Madison Avenue, and their interests were similarly aligned: monopolizing utilities. Adams began an extensive trip through England and the Continent, where he created a symposium of experts called the International Niagara Commission, a broad group of hydropower specialists and engineers who were offered an award for the best and most feasible design for the project. The experts agreed that it could be done, though there remained uncertainty about the most untested element: could 130,000 volts of alternating current travel twenty-six miles to the city of Buffalo? Also largely untested was Tesla's AC motor, which, in Adams's opinion, was "still a prophesy, rather than a completely demonstrated reality." Much of the financial return, beyond heating and lighting, would hinge on its success.

Nevertheless, Morgan and Adams formed the Cataract Construction Company and the Niagara Falls Power Company with $2,630,000 of start-up capital from a syndicate of one hundred select friends, "one of the most powerful combinations of New York capitalists," according to Scottish engineer Professor George Forbes, who was soon given a supervisory role for the project. The final cost would quickly prove to be much greater, an investment risk on a scale that no local public utility could or should have undertaken. In the absence of any state or federal utility, there was no other way this enormous undertaking could have been capitalized without J. P. Morgan and his syndicate of magnates.

They were, in the words of one Westinghouse engineer, "confronting a problem without precedent in its magnitude and almost without parallel in its significance." He took pains to emphasize the collective knowledge that made this project possible. "America is in no position to claim exclusive credit.... Were it possible to trace to its true source each one of the great number of ideas embodied in the complete installation, it is probable that we should find nearly every civilized nation represented—England, America, Switzerland, France, Germany, Italy, some in greater degree, some in less, but all cooperating to achieve what is, beyond question, one of the most significant triumphs of nineteenth-century engineering skills."

Digging proceeded in Niagara well before the contract for the generators was decided. This was the golden era of the steam shovel—basically, a rail car with mounted steam engine and crane—with which more than 2,500 men were removing 600,000 tons of rock and dirt for the tunnel and the wheel pits. By early 1893, "about a mile and a half above the American Fall a canal has been dug out 500 feet wide, and 1,500 feet long, with a depth of 12 feet," wrote George Forbes. "Along the edge of this canal, wheel pits are being dug 160 feet deep, at the bottom of which turbines will be placed." Stanford White, the preeminent architect of the era, was recruited to build a "cathedral of power" to house the first three of the five dynamos high above the turbines, with a second powerhouse for the remaining two. Each turbine weighed eighty-five tons and delivered 5,000 horsepower, vastly more powerful than the giant Corliss engine being installed just then at the Chicago Exposition. But which turbines would be chosen, generating what electricity at what frequency?

That May the board of directors of Cataract finally announced its decision. It was irrevocably committed to generating polyphase alternating current, thus allowing its transmission to Buffalo, and forgoing any mechanical waterpower from their Niagara site. Adams invited six companies to bid for the job: three from Switzerland as well as Westinghouse, General Electric, and Thomson-Houston. After months of analyzing all the detailed submissions, Adams informed the bidders that none would be accepted; instead, Professor Forbes would design the system himself. This provoked considerable outrage from everyone involved.

But before long Adams was consulting with Tesla, who made it clear that Forbes could not possibly produce the required current without infringing on numerous Tesla patents and provoking a ceaseless drumbeat of lawsuits. It was obvious that Forbes had appropriated many elements from the Westinghouse-Tesla system (as had General Electric) without fully understanding many of the outstanding issues, including the frequency of the current. Once the Westinghouse engineers had compromised so that Tesla's motors could operate on the system—that is, when the mountain came to Mohammet—the Cataract contract was theirs.

But something larger was unfolding. Following Barings Bank, a number of other investment banks had imploded, and because of the wild speculation in railroad stocks, hundreds of miles of expensive, unused track had been laid. When expenses outstripped revenues, the Northern Pacific and the Union Pacific railroads failed, among other major lines, and a run on gold followed, until the hemorrhaging Treasury could no longer redeem paper currency. The Panic of 1893 decimated some 500 banks and 15,000 companies.

By this time Morgan's syndicate had invested $4,000,000 in the Niagara project, and the costs continued to climb. Most of these investors also owned General Electric, which was growing more fragile by the month as it lost market share to AC central stations and faced infringement suits from Westinghouse by the score. More than three hundred lawsuits were in progress between the two companies, costing each a million dollars a year in legal fees alone. Consolidation of the two companies would mean a tremendous savings, and would handily corner the market with a near monopoly; it seemed so profitable as to be inevitable. As rumors spread, Westinghouse shares plummeted. George Westinghouse retaliated in kind, with rumors that reduced GE's stock value. Hostilities only ceased two years later when GE capitulated and paid royalties to use Tesla's AC patents. GE then began specializing in building industrial AC-DC transformers, among other things.

Jonnes recounts how Cataract Construction had set aside 1,500 acres adjoining the power plant for industries that would avail themselves of the generated power. In fact it became clear that there was a pent-up demand waiting for the electricity in order to launch new industries and new methods of production. Foremost among these was the smelting of aluminum by electric current from the aluminum oxide in clay. This process was developed by Chester Martin Hall, whose Pittsburgh Reduction Company was soon renamed the Aluminum Company of America, with the acronym Alcoa. Even today, Hall's electrometallurgical process remains the only viable method for extracting aluminum, and Alcoa remains among the largest consumers of electrical power in New York State. Since the 1950s, its destiny has been closely tied to the development of the New York Power Authority.

A variety of electrochemical processes that were unaffordable outside a laboratory were suddenly made feasible on an industrial scale. As well, many new products, like the abrasive carborundum, were being manufactured *because* electricity was available, and producers of "acetylene, alkalis, sodium, bleaches, caustic soda, chlorine" were vying for voltage, according to Niagara historian Pierre Berton. In fact, all 10,000 horsepower produced by the first two turbines were promptly taken up, along with nearby land—a welcome surprise to Adams and the investors. On August 26, 1895, the electricity from one dynamo (and days later from the second) began flowing. That much generation was fully spoken for by businesses just in Niagara Falls, where demand was growing for another 5,000 horsepower. As Hall's Pittsburgh Reduction Plant received the first power, the irony was apparent in a *New York Times* article. "The current sent is an alternating one, and before it can be used in the making of aluminum it must be transformed to a direct current. This is done by passing through four of the largest rotary transformers ever built. These are 2,100 horse-power each, and three of them are running." After all the additional cost, delay, and risk that alternating current had entailed for Cataract Construction, every one of its first customers required direct current.

But the accomplishment was no less impressive to the party of New York City investors who came to witness the unique enormity of their venture. Tesla also came at last to see what he had wrought, taking well-deserved pride in the stolid reality of what once had been his ephemeral vision on the banks of the Danube. And H. G. Wells, best remembered today for *The War of the Worlds* and other science fiction, was deeply moved by his visit. "These dynamos and turbines of the Niagara Falls Power Company impressed me far more profoundly than the Cave of the Winds.... They are *will* made visible, *thought* translated into easy and commanding things. They are clean, noiseless, starkly powerful.... These are altogether noble masses of machinery, huge black slumbering monsters." Of the hydropower, Wells wrote, "All the clatter and tumult of the early age of machinery is passed and gone here; there is no smoke, no cold grit, no dirt at all.... I fell into a daydream of the coming power of men, and how that power may be used by them."

The ramshackle tourist town was soon surrounded by a heavy industrial region just out of sight from the Falls. Within a decade a larger Schoellkopf powerhouse was added, with an output of 34,000 horsepower to serve the rapidly expanding industrial center of Niagara Falls and environs, demonstrating again that providing power attracted businesses and created jobs, dramatically increasing the productivity of small companies and entire regions. For a time, Niagara Falls was producing 20 percent of the electricity in the United States.

Electricity and the Transformation of America

The full value of alternating current would not be appreciated for another year, while the city fathers of Buffalo dickered about granting the franchise, how much electricity they would buy, and to what purpose. There was even some talk of a public utility similar to the one that Jamestown, New York, had created in 1891. Adams had required the city to commit to purchasing 10,000 horsepower before Cataract would begin digging for the three additional turbines and dynamos, and that agreement was finally signed a month after the first 1,000 horsepower were transmitted to the Buffalo trolley system. As Jill Jonnes elegantly describes, the AC electricity "flashed out at a pressure of 2,200 volts, being swiftly stepped up in the GE transformer to 10,700 volts, flashed over the 26 miles of cable to the Cataract Power and Conduit Company's transformer.... The Buffalo streetcars were soon jogging along their rails with Niagara hydropower, mundane and yet wondrous proof of this miraculous new and invisible reality—long-distance transmission of AC energy." "Electrical experts say the time [it took] was incapable of computation," reported the *Buffalo Enquirer,* poetically describing the event as "the journey of God's own lightning bound over to the employ of man."

The "Queen City of the Lakes" had entered the Electric Age, thanks to Tesla, Westinghouse, and, of course, Morgan's syndicate. Their total investment had risen to six million dollars, a great sum to place at risk—so great that several carped angrily and publicly about it. But what their money bought was a natural monopoly for years to come: the overwhelming cost advantage and pricing power of a single supplier to a population of a quarter of a million people in a major metropolis. And of course, unlike most power companies, they did not have the cost of coal and its transport to pay year in and year out. Buffalo, the western terminus of the Erie Canal, was already the eighth largest city in the United States, and New York's second largest. A major central rail hub, its silos stored much of the grain from the Midwest traveling east on the Canal, making this a dynamic, forward-looking metropolis. The largest office building in the world opened in downtown Buffalo. The electric streetcars joined the national trend away from dangerous and unreliable steam-powered trolleys. And unlike every other city with power, Buffalo was not polluting itself with coal-fired generation.

In Buffalo, as in most newly wired cities, the luxury of electricity soon seemed conventional before it became a necessity, and in no time "the Queen City" was known as "the City of Light." The brightness of streetlights was often superseded by lighted advertising, and not only on in New York City, where one dynamic, lighted sign stood 72 feet high and 90 feet wide. Small towns from New York to the Midwest created their own White Ways as displays of conspicuous consumption. Lighting the shops and streets from the early hours of dusk transformed cities, suggesting a sense of the possible glamour and excitement of nightlife in even the dullest small town. Lighting "drew more trade to their stores," observed social historian David Nye, "improved traffic conditions, deterred crime, and boosted civic pride."

Nye also noted that "electrification displaced workers, yet it did not create unemployment because it fostered new businesses.... [It had] multiple effects, transforming older production methods, de-skilling some jobs, eliminating others." As more businesses subscribed to electricity, more and more cable was laid down, often through residential areas, creating much greater domestic access. This was ideal for the utilities, which could then provide the bulk of their power to industries during the workday and to residences at night. Across the country, most utilities charged a flat rate to domestic customers but supplied only 100 watts, useful only for lighting.

By 1905, flat rates were disappearing with the widespread introduction of electricity meters. Thereupon, all sorts of new household appliances were quickly invented and marketed: hot plates for cooking, irons, toasters, popcorn poppers, percolators, heating pads, water heaters, washing machines, and vacuum cleaners. Electric fans in the summer seemed almost magical, but nowhere near as valuable as electric heaters in winter. Once Frigidaire was established in 1916, electric refrigerators would be mass-produced, though ice boxes remained cheaper and more prevalent for years. With the introduction of the radio in the early 1920s—thanks largely to the later work of Nikola Tesla—the entire world of news and entertainment was revolutionized in urban communities by live music, news bulletins, and the first soap operas. Of course the vast majority of farms would only be wired up after 1935, thanks to the New Deal's Rural Electrification Administration.

Meanwhile, across urban America, the demand for electricity multiplied year after year. Its many new uses in homes, workplaces, and public venues were transformative to the culture as a whole. Commerce expanded into the night hours, while advertising and shop windows were often lit night and day. Technical schools that sprang up to train engineers and electrical apprentices could now offer night classes. Electrical repair shops proliferated. Homeowners who replaced gas jets with sockets paid less for fire insurance. By 1900, in many cities you could see a distinct dividing line between the smoky, dingy neighborhoods reeking of kerosene and the aspirational modernity of well-lit, residential areas. Absent any regulatory authority, electrification bypassed whole regions that could not produce a profit for investor-owned utilities.

Of course, electricity was the theme of the Pan-American Exposition held in Buffalo in 1901, the next great demonstration of its wonderments made possible by the world's largest power station at Niagara twenty-five miles away, proving to one and all the reliability of long-range transmission. The exhibit was illuminated by no less than 200,000 small incandescent bulbs rather than a score of arc lamps, and included a scale model of Niagara Falls, along with a variety of extravagant electric amusement rides not unlike today's, including "Aquarama" and "A Trip to the Moon." The new GE X-ray machine was showcased; though when President McKinley was shot at the Exposition in September, doctors did not dare use it to locate the bullets in his abdomen. His death after a week threw the nation and the stock market into shock, as Vice-President Theodore Roosevelt took his place. Nonetheless, some eight million visitors attended the Exposition in Buffalo in all and witnessed the power of hydroelectric generation.

Hydropower in the Future

In 1900, the level of carbon dioxide (CO_2) in the Earth's atmosphere was largely unchanged since preindustrial times: approximately 278 parts per million. That same year, the human population reached a record of some 1.6 billion; for 99 percent of our species' history, the population had remained below 200 million.

By 2012, atmospheric carbon dioxide had reached a peak of 393 parts per million as the global population approached seven billion. According to renowned climatologist James Hansen and colleagues, "a CO_2 amount of order 450 ppm or larger, if long maintained, would push Earth toward the ice-free state. Although ocean and ice sheet inertia limit the rate of climate change, such a CO_2 level likely would cause the passing of climate tipping points and initiate dynamic responses that could be out of humanity's control." Hansen, whose congressional testimony on climate change in 1988 helped launch the Intergovernmental Panel on Climate Change (IPCC), to which President George H.W. Bush made important contributions, adds an ominous reminder: "The climate system, because of its inertia, has not yet fully responded to the recent increase of human-made climate forcings." Since CO_2 persists for many decades, the greatest amount of atmospheric CO_2 today is still the result of England's long history, going back to the nineteenth century, of burning coal for home heating as well as industry. Today, the People's Republic of China has surpassed the United States in producing the largest amount of CO_2, closely followed by India and Russia. The single greatest use of coal today, by far, is for the generation of electric power.

The solution to this problem was demonstrated more than a century ago with the renewable energy launched at Niagara Falls.

Beyond Lighting

By 1905 George Westinghouse had moved on to the first full-sized electric locomotives, which he operated on AC near his Pittsburgh plant, and which was soon adopted by the New York, New Haven, and Hartford Railroads. But just as his production expanded and needed capital, a fresh scandal arose on Wall Street that once again involved an attempt to corner shares in a Montana company, United Copper. When this attempt failed, it triggered the collapse of several associated banks, and alarm turned into the financial hysteria known as the Panic of 1907. By October, after Westinghouse failed to raise the capital needed to deliver on his locomotive orders, his company went into bankruptcy. When the company came out of receivership the following year, banking interests had majority control and the new corporate president was a lawyer chosen by the bankers. Two years later, George Westinghouse was removed altogether from the board of the Westinghouse Electric and Manufacturing Company, "a disappointment from which he never recovered," said a colleague. "There is no doubt that it broke his spirit."

At a gathering of the Society of Mechanical Engineers in 1911, Westinghouse spoke out about the damaging impact of unregulated capitalism upon the creation and development of new technologies and the threat of financial syndicates, or "trusts," as they were known. "Their very magnitude," he declared, "coupled with the evil practices so frequently disclosed in press and in our law courts, has so aroused the public that there is now a fixed determination to establish by national and state laws an exacting governmental control of practically all forms of corporation, in order that competition may be encouraged and not stifled, but seemingly with due regard to the real objects in view—the securing of the best public service in all forms.... But fortunately there are indications that the great leaders are alive to the importance of the regulation of legislation."

The "great leaders" he referred to were the progressive Republicans he supported: Theodore Roosevelt, President William Howard Taft, and Charles Evans Hughes.

CHAPTER FOUR

THE NEED FOR PUBLIC POWER

Luxury or Necessity?

In the early 1900s, New York State vividly illustrated why the regulation of privately owned utilities was essential and public power indispensable. Much as rivers in the state were providing the potential energy that was transformed into electricity, that electric current was providing the potential energy for social transformation. Generated power increased productivity so dramatically that its introduction into a community created jobs, growth, and wealth. Conversely, communities that failed to modernize with electricity quickly lagged behind, triggering a downward spiral for jobs, population, tax base, and municipal services. This tailspin was repeated across cities and regions, demonstrating undeniably that power had become a necessity for a neighborhood or town just to hold its own. It was clear now that privately owned companies determined the future of whole communities and even regions, picking winners and losers on the basis of profits anticipated and, all too often, bribes received.

The new technology raised new questions: is electricity a luxury? Or is it, like water, a necessary service that state and local governments should guarantee? If not, J. P. Morgan was free to determine the price of electricity as he saw fit, and once he merged Edison and several other companies into the Consolidated Gas Company, his monopoly over energy, aided by the political corruption permeating every layer of government, gave a handful of well-dressed, like-minded men near-absolute control over the daily life and long-term destiny of almost every community.

To understand the helplessness of neighborhoods that could not attract power companies, it may be helpful to recognize that today, "the dispute over municipal broadband bears a striking similarity to the development of the electric power industry a century ago," as Robert McChesney and John Podesta noted in "Let There Be Wi-Fi" That similarity, according to public attorney James Baller, "casts substantial doubt on the notion that our nation can depend on competition among cable and telephone companies alone" to deliver universal, affordable access to the information superhighway. Baller showed that

> when electricity first became available in the 1880s, privately owned utilities marketed "the new technology as synonymous with wealth, power and privilege," lighting large cities, businesses, and the homes of the rich. Electricity also allowed factories to stay open twenty-four hours a day and led to the institution of swing shifts. But communities that didn't have electricity couldn't produce as much, and couldn't keep up with urban competitors. Rural communities were left with the choice of forming a government-owned utility or being left in the dark. Even big cities like Detroit built municipal power systems to cut prices and extend service. In response, private utility companies launched a massive propaganda and misinformation campaign that attacked advocates of municipal power as "un-American," "Bolshevik," and "an unholy alliance of radicals."

Although the language is less inflammatory today, advocates of "net neutrality" and government-supported broadband in rural and disadvantaged communities face the same free-market arguments. In Europe, where the public sector is more robust, Wi-Fi has been widely available for years in public places (airports, railway cars, and subways), either for free or for a modest fee; others offer it as a public service at taxpayers' expense. But even though such widespread connectivity would benefit millions of Americans (on commuter trains, for example), that is unlikely to happen without significant legislative reform. Similarly, in the 1890s, nothing but the determination of reform-minded legislators could curtail exploitation and discrimination by power monopolies. It took more than a decade to marshal that determination, during which political momentum swung to and fro almost annually between reform and corruption of every kind.

Theodore Roosevelt and the New York Transformation

Reform began from the bottom up. In 1892, Presbyterian minister Charles H. Parkhurst set out on a "Sin Tour" to visit the lowest of Manhattan's thousands of low-life bars and bordellos. Guided by a "sporting man" and a former private detective, Parkhurst traveled from the Tenderloin brothels near his Madison Square church to downtown opium dens and the cheapest "penny hangs," where the whiskey was five cents a glass and tasted like "kerosene oil, soft soap, alcohol and the chemicals used in fire extinguishers." They descended from the "French Circus" on West 4th Street to the Golden Rule Pleasure Club, where boys in make-up with "high falsetto voices" were prostituted. All these illegal establishments and thousands more were left undisturbed by policemen who received scheduled bribes and in turn paid graft to their superior officers, who surrendered a percentage of their take to the ward bosses and bagmen sent by Tammany Hall. Thus a chain of corruption stretched from the gutters of the Tenderloin to the highest offices in city and state governments: bribes collected from the penny hangs helped finance the reelection campaigns of politicians who served the men who invested in utilities.

The following Sunday, Reverend Parkhurst gave a sermon entitled "The Wicked Walk on Every Side When the Vilest Men are Exalted." He described "going down into the disgusting depths of this Tammany-debauched town and it is rotten with a rottenness that is unspeakable and indescribable.... To say that the police do not know what is going on ... is rot.... Drunkness, gambling and licentiousness in this town are municipally protected."

That was because, although Boss Tweed himself was long gone, Tweed's system and the Democratic political machine that he had built at Tammany Hall remained "strategically deployed," according to historian Kenneth Ackerman, "to control key power points: the courts, the legislature, the treasury and the ballot box. Its frauds had a grandeur of scale and an elegance of structure: money laundering, profit sharing and organization." It is difficult to imagine how such a vast criminal conspiracy endured for decades, but as Ackerman explains:

> Tweed's system had an irresistible political logic: everyone benefited, rich and poor alike, and nobody seemed to get hurt. Money for graft as well as good came mostly from outsiders. Taxes stayed low; [the city comptroller, Richard "Slippery Dick"] Connolly financed city operations mostly with debt, selling bonds and stock to investors in Europe and on Wall Street, pushing off payment until another day. For the wealthy, Tweed produced dynamism and growth. His regime spent $10.4 million on Central Park up through 1869, and millions afterward. Wasteful and riddled with graft? Certainly. But during this time, the value of real estate in the three surrounding wards more than tripled.... The problem was, Tweed's system was based on lies.... Eventually, the bills came due, the creditors got nervous, and the house of cards collapsed. Meanwhile, democracy itself almost drowned in the process: self-government meant little when elections were won by the side that cheated best.

Tammany Hall, which was effectively the headquarters of the state's Democratic Party, focused on recruiting the escalating numbers of New York City's working-class and fresh immigrants, while Republican legislators, led by the Wall Street cabal, controlled most of the state. Since their profit centers and territory were separate, the two established a *modus vivendi* outside of election season, almost complementing each other like corresponding cogs in a gear. To keep that machine well-oiled, Tammany leaders always kept a few Republicans on the city payroll. But mutual cooperation was almost assured by the methods they shared, since both parties were, in the words of one early reformer, "representatives of the spoils system; both are absolutely opposed to nonpartisanship in municipal affairs; both believe that this city is a rich field to be harvested only by party spoils men; both represent a gang of plunderers who seek to prey upon the prosperity of the city." And both sides understood that if you "Steal a little, they throw you in jail, / Steal a lot and they make you king," as Bob Dylan put it laconically.

Republicans who safeguarded their own financial interests in the state legislature were dubbed "the Plunderbund," defined as "a league of commercial, political, or financial interests that exploits the public." Their key interest in Albany was to impede *any* effective regulation or taxation of the business sectors they monopolized, specifically, railroads, insurance, gas, and electricity; in short, the realm of the robber barons. For many years they met very little political resistance from the citizenry, who seemed content to believe what they read in the well-bribed press. Political historian Robert E. Wesser noted the "deep gulf between those who enjoyed the public trust and their complacent, often indifferent constituencies." As one leader in Albany explained, "the populace had to be cajoled, educated, flattered, and given candy, 'or its judgment will not be so good.'"

Parkhurst's foray into the sinful soul of the city produced immediate counter-charges, including testimony that he had himself played leap frog with a group of brothel girls. But it also led to the trial of several police officers, and it was their acquittal by crooked judges that aroused public indignation at last. Over the next two years a special commission dug deep and cleaned house, charging fifty-three officers, including current and former commissioners, with corruption or incompetence, while forcing many others to resign. That gave reformers the upper hand and an empty roster, so in 1895 a newly elected Republican mayor surprised everyone by appointing a little-known figure to preside over the city's

Police Commissioners: a progressive Republican named Theodore Roosevelt.

The future president's two-year stint as Police Commissioner is succinctly summarized in the title of Richard Zacks's *Island of Vice: Theodore Roosevelt's Doomed Quest to Clean Up Sin-Loving New York.* According to the New York Police Department's History Division, "An iron-willed leader of unimpeachable honesty, Theodore Roosevelt brought a reforming zeal" and "set the standard for the modern NYPD." On his watch, 1,600 new recruits were signed up without regard to their political affiliation, and the first ethnic minorities and even a woman were hired. There is no doubt that TR, as he was often called (never "Teddy"), first made his national reputation as a Republican reformer during this time, though the NYPD acknowledges only that "some of his reforms were undermined by later Tammany regimes"—a mild understatement.

Most New Yorkers first noticed TR's zeal when he began enforcing long-ignored "Sabbath laws" that forbade the sale of alcohol on Sundays. At first that brought him widespread, if lukewarm, praise. But as soon as the public turned on him, the press labeled him "a little tin Czar." To avoid the enforced abstinence, on Sundays an estimated 25 percent of the city's population fled "Roosevelt's Deserted Island" (as the *Herald* described Manhattan) to drink in New Jersey or Connecticut, or in upstate towns like Albany where the alcohol always flowed freely, state law notwithstanding. Millions in revenue were lost, hundreds of brewers and bartenders laid off, and a quarter of New York City's 2,000 saloons faced foreclosure.

TR did not budge. In fact, he doubled down, ordering raids on well-known brothels frequented by well-to-do "gentlemen," and having dozens of saloonkeepers arrested every Sunday. What drove him in his zealous crusade? Throughout his life, Roosevelt demonstrated a genuine and consistent puritanical streak; but it is also worth noting, as Zacks does, that just a year earlier he had lost his beloved brother Elliott to a long, horrific struggle with alcoholism and behavior so egregious that it had threatened their vast family fortune. And, like Parkhurst, TR had witnessed with disgust the seediest parts of the city on midnight rambles, though unlike the pastor, he was there to check up on the cops, who were often nowhere to be found. From his puritanical, personal, and policy perspective, he was inflexibly committed to raising the police force to the highest standards of moral and professional conduct.

As petty a matter as public imbibing on Sundays might seem, this early Republican reform by TR triggered the backlash that swept Tammany Hall Democrats back into City Hall. Every community had its drink of choice, from the genteel wine-sipping WASPs to rough Irish tipplers; but it was the German beer drinkers who mustered the most political clout. Both political parties were internally divided over the issue, while most of the police force hated TR and the loss of bribes. The *New York World* quoted one cop complaining that "the law is no good anyhow. These poor people like to have a glass or two of beer on Sunday, which without it is no Sunday at all to them.... I guess this crusade is causing more suffering to poor families than anything else they have to bear." TR was also widely attacked for enforcing the laws selectively, leaving the gentlemen's clubs unaffected and similar inequities. Most of his political trouble came from Thomas Platt, the longtime political boss of old-guard, *anti*-reform Republicans, who was prepared to accept most anything that pleased his Wall Street backers. Platt's political motto was said to be: "Whoever pushes you up the tree is the boy to share the apples with."

Republican mayor Seth Low, wary of firing the moral crusader outright, instead threatened to eliminate the entire Board of Police Commissioners. So in the end, after winning a few important skirmishes, Roosevelt lost his war on corruption, defeated by unrepentant Tammany Democrats, scorned by ungrateful old-guard Republicans, and stymied by his fellow Police Commissioners. As TR wrote to a friend, he had some support from those who "make up the spine of the Republican party, but no hold whatever on the people who run the Republican machine," referring to Thomas Platt *et al.* After less than two years, TR urgently sought a graceful exit from his nearly ruinous foray into city politics. His friends, concerned that this episode could terminate his political career, scrambled to find him a suitable federal post; he was saved when President McKinley offered him the position of Assistant

Secretary of the Navy. He left that office to create the "Rough Riders" Voluntary Cavalry Brigade, which he famously led into battle against Spanish troops in Cuba, earning a stellar reputation nationally.

But TR also had the last laugh in New York. By 1898, as the public demanded reform, Republicans were desperate for a squeaky-clean candidate for governor, and no one fit that description better than the crusading former Police Commissioner. Thanks to his war record, Roosevelt won the election, although by just one percent. When he completed his two-year term (as the governorship was then), he had achieved only modest reforms against unceasing Wall Street resistance, but left office with a greatly enhanced résumé. He had become a shrewd political player, even neutralizing the magnates' clout for a time. Platt saw that his best chance of eliminating TR from the New York scene was to help him become President McKinley's running mate in 1900. This strategy was successful, but only in the short term. In fact, perhaps the most enduring outcome of Roosevelt's governorship was to make him a wily, knowledgeable, and admired force in New York politics, perfectly placed in the White House after McKinley's assassination, to support progressive Republicans as they fundamentally reformed the state's power industry.

By 1905, the year after Roosevelt's election as President in his own right, Consolidated Gas had accumulated capital assets exceeding $80 million even after buying out every competitor in the vicinity of New York City. It had also expanded by buying up several small electric companies. Among its largest customers was the city itself, which purchased electricity for the streetlights of Manhattan and the Bronx at whatever price Consolidated dictated. The law required that the price be fixed by public bidding for the contract, but in the absence of any competitors, that point was moot. Unable to force the company to negotiate its prices with New York City instead of dictating them, the new Democratic mayor, George B. McClennan, Jr., proposed the creation of a city-owned power plant for the streetlights as some in his party had before. Platt and the Republicans vetoed McClennan's proposal, producing real, popular outrage, leading to the Stevens Commission, the first state investigation into utility and insurance pricing. To conduct the hearings, the Commission recruited Charles Evans Hughes as its legal counsel.

The Stevens Commission

Hughes was born in Glens Falls and earned his law degree at Columbia Law School before joining a major law practice. A lifelong Republican with no political ambition, he was 43 years old when appointed counsel for the Stevens Commission in 1905. From the beginning, his diligence in preparation and meticulous questioning of witnesses set a new standard for state commissions in general, and his fair-mindedness earned the grudging respect of his worst enemies, including party boss Thomas Platt and the political clique surrounding J. P. Morgan, who all saw Hughes as a Republican turncoat.

As the Stevens Commission went to work, New York joined the fierce power wars. There were some 2,500 private power companies around the country, all of which were controlled to some degree by Morgan and a relative handful of investors. There were also more than 800 municipal power companies, with 60 to 120 new public plants built each year between 1897 and 1907. Most of these were in small towns for which public ownership was the only option, but according to the American Public Power Association, "nonprofit local ownership meant operating costs could be about half of private costs," highlighting the exorbitant pricing by private utilities everywhere else. So wherever General Electric—which had been bought by Consolidated Gas—chose to operate, it pulled every political lever it controlled to prevent the creation of municipal power plants; since the bribes that pulled those levers amounted to less than half the operating costs, it was clearly worth the cost.

As chronicled by the authors of *Power Struggle: The Hundred-Year War Over Electricity*, evidence nationwide showed "a steady stream of city councilors for sale and government officials acting on behalf of private interests.... Cleveland [Ohio] came to embody the early struggles over electric power that also emerged in San Francisco, Chicago, Detroit, and hundreds of smaller cities." Three months before the Stevens hearings began, Cleveland's Mayor Tom Johnson set an example that reverberated across the country. He accused seventeen city council members of receiving large campaign donations or outright bribes from the Cleveland Electric Lighting Co. (CELC), owned by GE. "I believe in municipal ownership of these monopolies," declared Mayor Johnson, "because if you do not own them, they will in time own you. They will destroy your politics, corrupt your institutions and, finally, destroy your liberties." The city fought J. P. Morgan for more than a decade before it succeeded in establishing a publicly owned power company, which charged three cents per kilowatt-hour, compared to the inflated ten cents charged by CELC.

That was exactly the kind of price manipulation that Hughes exposed in April 1905 with the Stevens Commission, as he publicly interrogated various executives of Consolidated, former commissioners, and Tammany chiefs. He had spent weeks in preparation, studying the accounts of various companies and uncovering how they consistently inflated the costs of production and thereby reduced their visible profits. Most important, wrote his biographer:

> were the revelations affecting the huge holding company at the apex of the utility pyramid. Hughes carefully analyzed the successive stages through which Consolidated Gas had moved in its drive toward monopoly. He thus learned that the company was overcapitalized in the amount of almost $8,000,000.... Hughes could find no justification for what he regarded as a form of extortion. When company officials tried to pass off this $8,000,000 sum as "good will," [Hughes] sardonically replied that in a monopolistic situation there could be no true criteria by which to judge good will....

From GE's records, Hughes's careful calculations showed that the actual cost of producing one kilowatt-hour of electricity came to 3.66 cents, while New York consumers paid an average 12.27 cents, yielding almost 300 percent profit for Morgan's Consolidated Gas. As consumer outrage grew, so did a call for serious reform and regulation, along with the belief that Charles Evans Hughes was just the man to carry it out, despite his firm declaration that he had no interest in becoming a politician, much less the next governor of New York.

It was President Roosevelt who convinced him of his civic obligation to run because, he wrote to Hughes, "You believe in reforming the relations between the government and the great corporations as drastically as is necessary to meet the needs of the situation, but you believe in having it done in a spirit of sanity and justice." As Wesser wrote, Hughes "forthrightly championed the Roosevelt theory of regulation, a theory basic to his own gas and electric commission bill.... The chief difficulty, he insisted, was not that the corporation per se was bad, but that in so many enterprises there was an 'evasion of responsibility.'" At this stage, "Hughes hoped to make public service corporations more responsible to the citizenry without resort[ing] to municipal or state ownership." Even before his election he proposed a new Commission of Gas and Electricity, which the legislature later adopted.

The Democrat running against him, newspaper publisher William Randolph Hearst, went further and called for a municipal power system for New York City. But Hughes held to his belief that "there should be effective governmental control of all great enterprises in which the public is interested," companies that "depend upon the public for their right to be and to do" because they require state and city charters or licenses. He consistently hewed close to "the basic philosophy of Rooseveltian reform," wrote Wesser, by insisting that "the Republican Party must adjust to changing times. The moral precepts of American civilization, now clearly audible in the voice of protest, must be reinstated as the sole impulse to political action, thus terminating the evil politico-business alliance especially evident on the state and local levels."

Hughes also vigorously asserted the state's "capacity to control and regulate these great enterprises, despite the influence which they are able to exert and despite the conflict in many ways between public and private interests." As for the Republicans' free-market complaint that government was becoming "too paternal," Hughes replied, "Let not those who enjoy the benefits of paternalism in the creation of corporations object to paternalism in their just regulation in the public interest."

The Legacy of Governor Hughes

Hughes was elected governor in November 1906 "by an eruption of respectability, with the negative aid of the most arrogant exponent of the corruption he denounced," according to one scholar. While every other Republican candidate statewide lost, Hughes won because of the integrity he demonstrated and the conclusions he drew during his investigation of the utilities, railroad, and insurance industries. "My feelings," he told the *New York Times* the next day, "are not those of elation, but those of responsibility." The same day, President Roosevelt wrote, "I feel that you and I, my dear Mr. Hughes, approach our work in the same spirit," and that was the clear-eyed spirit of political independence. Hughes's inaugural speech and his first speech to the legislature were described by the press as "an open challenge to the bosses of his own party," striking directly at corruption and delivered "in the interest of the people by a man who seems to be their, not the politicians', servant," which would "disjoint the old Republican machine."

Two months later Hughes presented his plan for the Public Service Law to regulate electricity, gas, railroads, and insurance. It called for *two* commissions, five members each, one for New York City and another supervising the rest of the state, where different issues and values prevailed. Both commissions would have the power "to pass upon the issue of stocks and bonds; to examine properties, books and accounts; to require detailed reports in prescribed form; to prescribe reasonable rates; to require adequate and impartial service; to provide for the safety of employees and for the protection of the public ... and generally to direct whatever may be necessary and proper ... to secure the fulfillment of the public obligation of the corporations under its supervision." It is safe to say that the notion of "the public obligation of the corporations" was utterly foreign to J. P. Morgan *et al.*, who were gobsmacked by, as they saw it, such "usurpation of power," "confiscation," and "ruination of our prosperity."

This bill (and a similar one in Wisconsin, from which some language was borrowed) represented the first meaningful effort at state regulation of public service corporations, as the state defined them (excluding the telephone and telegraph). A primary goal of the legislation, according to an accounting specialist, "was to ensure the 'integrity of capital' and the correctness of the charges to 'cost of operations.... our greatest instrument of regulatory control.'" Hughes had already shown New Yorkers exactly how the utilities—not satisfied with wildly overcharging—cooked their books: by vastly exaggerating and falsifying their costs of doing business, they minimized their apparent profits, with the added benefit of reducing their corporate taxes, local, state, and federal. Yes, *of course* this was illegal, but New York had never yet had an untainted, independent authority that could possibly prosecute the financiers. Now that was going to change.

The bill aimed to shield the two Public Service Commissions from interference by spurious legal appeals, lest "the court becomes in effect the ruling commission." Judges, Hughes argued, lack the necessary special knowledge and experience to understand the complexity of power generation (or insurance, railroads, gas, etc.). Each commission would have "power to act upon its own initiative as well as upon complaint; to pass upon the issue of stocks and bonds; to examine properties, books and accounts; to require detailed reports in prescribed form; to prescribe reasonable rates; to require adequate and impartial service; to provide for the safety of employees and for the protection of the public ..." The bill also gave the governor the authority to replace commission members. In fact, the law that Hughes proposed has served as a blueprint for the even-handed regulation of utilities ever since.

Before the bill was voted on, Hughes traveled around the state to introduce himself as governor and, perhaps more importantly, to put public pressure on Albany to pass the bill by going "over the heads" of the corrupt politicians to appeal directly to the people. In two months he spoke at forty dinners and meetings, from Long Island to Buffalo. His speech in Elmira was an eloquent articulation of his credo.

> I am here under a retainer. I am retained by the people of the state of New York, to see that justice is done, and with no disposition to injure any investment, but with every desire to give the fullest opportunity to enterprise.... [What the people] revolt against is dishonest finance. What they are in rebellion against is favoritism which gives a chance to one man to move his goods and not to another; which gives to one man one set of terms and another set to his rival; which makes one man rich and drives another man into bankruptcy or into combinations with his more successful competitor. It is a revolt against all the influences which have grown out of an unlicensed freedom, and of a failure to recognize that *these great privileges, so necessary for public welfare, have been created by the public for the public benefit,* and not primarily for private advantage.

By June 1907, Hughes had succeeded in convincing the people at large with his well-spoken argument. When one state senator got home from Albany, he wrote, "the people fairly mobbed me." Within days Republicans were rushing back to Albany from their vacation to support the governor's plan *unanimously* and, in the words of one commentator, "to save the Republican party from suicide"; another opined that "an insolent and hostile legislature was on its knees." That June, when the Public Service Law was enacted, *Pearson's Magazine* pronounced it "the most remarkable and, in its significance, far-reaching law passed by any state in the last quarter of a century, a law that may profoundly affect the whole future of the nation."

Sustainability in the Public Interest

Another progressive concern that Governor Hughes shared with President Roosevelt was conservation of the state's natural resources and environment—sustainability, in today's parlance. Until then, the legislature routinely granted charters to public lands for the private benefit of corporations, usually in perpetuity and at no cost. Hughes found that unacceptable, and in his first year in office laid down this guiding principle: "These resources should be preserved and held for the benefit of the people and should not be surrendered to private interests." Explicitly included among those resources was the potential for hydroelectric generation. So when the Long Sault Development Co. sought a routine charter to build hydropower dams along part of the St. Lawrence River, Hughes surprised them by hiring an expert to estimate their potential profits and requiring a payment of $100,000 up front and annual rent thereafter. This assertion of the state's right to the usufruct of its lands set an important precedent nationally.

Hughes attended a governor's conference at the White House the following year, where his speech rallied the conservation movement championed by President Roosevelt. He acknowledged that New York was wasting its potential of 550,000 horsepower of energy from hydropower, worth $6,600,000 each year, because it was only now developing a comprehensive policy for state-owned generation. He urged the other governors to explore the hydropower potential in their own states and devise their own plans. Such competition from the public sector was the only real threat to investor-owned utilities across the country.

In 1910, his last year in office, Hughes had enough data to formulate a comprehensive policy for the systematic development of hydroelectric power along New York's major rivers. That policy was based upon five precepts:

1) River flow should be regulated and water power developed "to the fullest extent that may be practicable";
2) Only the state should build and own regulatory or power-generating reservoirs on streams originating in or flowing through public parks or state reservations;
3) Hydroelectric power projects should be undertaken by the state "whenever such action appears to be feasible and for the general interest";
4) Power so developed should be made available on "equitable terms and conditions";

5) Rigid scrutiny should be provided to see that the public interest in the state forest preserves would not be jeopardized by water-power projects.

Those precepts remain at the heart of the New York Power Authority today, though hardly anyone knows that they originated more than a century ago. Hughes's comprehensive policy that November was not lost on one newly minted Democratic state senator: Franklin Delano Roosevelt had just won his first election. The 28-year-old absorbed the precepts Hughes put forward. Later, he would put them into action, from the Power Authority in New York, to the Tennessee Valley Authority, to Bonneville and Grand Coulee in Oregon and Washington.

CHAPTER FIVE

FDR CREATES THE POWER AUTHORITY

FDR Enters Politics

Franklin D. Roosevelt, born and raised beside the Hudson River in Dutchess County, had only developed an interest in politics after getting to know President Theodore Roosevelt, his wife Eleanor's uncle and his own distant cousin. In 1905, when the fledgling lawyer married Eleanor Roosevelt, her uncle TR came from the White House to give his niece away. Their subsequent conversations, and Franklin's visits to the White House during TR's presidency, goaded him to engage in the politics of his county and then his state.

In 1910, Hughes resigned as governor when he was appointed to the US Supreme Court by TR's Republican successor, President William Howard Taft, thanks largely to his regulatory impact—not generally a Republican forte. State elections were approaching that year, and though the counties along the Hudson were solidly Republican like most of rural New York, the city of Poughkeepsie, near FDR's Hyde Park home, was Democratic. The political leaders there expected to lose the state senate race, but they wanted to make a good showing, and after meeting FDR, they offered him their party's nomination. Since he could not see any benefit to losing as he was expected to do, he decided to win. He set out on one of the world's first campaigns by automobile, driving across Columbia, Dutchess, and Putnam counties and reaching out personally to the farming population like no candidate before him.

As it happened, the election of 1910 was a Democratic sweep, because the Republicans were so divided over the person and policies of TR that many rejected his chosen candidates. For the first time in decades, the Democrats had control of both houses in Albany. "Mr. Roosevelt carried Dutchess County triumphantly," wrote the Poughkeepsie *News-Press*, "and was unexpectedly strong in the rural districts where it was supposed that he would lose heavily." He was only the second Democrat to win his district in fifty years.

Franklin and Eleanor moved to Albany so that Franklin could devote himself to this new career, which began with a three-month battle against Tammany Hall over their choice for the next Democratic nominee for the US Senate, "Blue-eyed Billy" Sheehan. Fighting the entire political machine of his own party, FDR gradually won allies and public support. He was not yet thirty years old, and the *New York Times* observed, he was "tall and lithe. With his handsome face and his form of supple strength he could make a fortune on the stage.... But no one would suspect behind that highly polished exterior the quiet force and determination that now are sending shivers down the spine of Tammany's striped mascot."

The two years that FDR spent in the state legislature established his reputation throughout the state and beyond as an insurgent against the Democratic political machine in Albany and in the national Democratic Party. His "insurgency" to block the Sheehan nomination paralyzed the legislature for two months; against all odds, he won that battle. "From the ruins of the political machines," he crowed during a speech in Buffalo, "we will reconstruct something more nearly conforming to a democratic conception of government."

But then the junior state legislator went much further with a populist demand for the direct election of US senators by the citizenry. As originally decreed in the US Constitution, the states' legislatures appointed their US senators, which allowed party bosses to handpick their members of the Senate. The young politician led the direct-election campaign at the national level, and in 1913 three-quarters of the US state legislatures agreed with him by ratifying the Seventeenth Amendment. For a political neophyte, this was an unparalleled achievement that would be remembered around the country; so would his early support for the women's suffrage movement, which produced the Nineteenth Amendment in 1920.

FDR was also the first New York supporter of Woodrow Wilson's campaign for the Presidency in 1912. He traveled the state, speaking on Wilson's behalf and thwarting Tammany Hall's machinations for Taft as well as the ambitions of Theodore Roosevelt, whose departure from the Republican Party and formation of the Progressive "Bull Moose" Party succeeded only in siphoning votes from Taft's reelection campaign and failed to give him even New York's voters. That party rift delivered the election to Democrat Wilson, who soon after appointed FDR as Assistant Secretary of the Navy.

During his seven years in that post, FDR was deeply involved in the country's strategic and tactical naval decisions of World War I, from which he unwittingly received a valuable education for World War II. To counter the unrelenting destruction of US shipping by German submarines, after the United States entered the war in 1917, FDR persuaded the US Admiralty to launch and support a project to deny the North Sea to the German' menace with a barrier of mines stretching 240 miles from the Orkney Islands to the Norwegian coast. The technology that he helped develop for the 56,000 mines in the "North Sea barrage" demoralized German submarine crews to the point of mutiny, and was believed by some to have accelerated Germany's capitulation in the Armistice the following year.

In 1920, after his run for the Vice-Presidency on James Cox's failed ticket against Warren Harding, FDR was stricken with the unimaginable horror of polio. Four years later friends urged his return to political life, but he was committed to advancing his recovery until the day he could walk. By 1928, after seven years of strenuous physical therapy, and using leg braces, he was able to stand up and shuffle awkwardly while leaning on the arm of one of his tall, strong sons, usually Franklin Roosevelt, Jr. All along, from his home in Hyde Park or the therapeutic resort in Warm Springs, Georgia, he had remained as politically engaged as ever. In New York politics, his ally was Governor Al Smith, who carried forward Hughes's call for public waterpower as far as he could before handing it over to FDR.

Introducing the Water Power Authority

Al Smith had served his twelve years in the Assembly as a useful mouthpiece and an unfailing cog in the Tammany Hall political machine. But after his first term as Governor (1918–1920), he showed more independence. Reformers like FDR saw that "Smith was not a typical politician," according to a contemporary, but that this "red-faced, jut-nosed, gold-toothed, harsh-voiced smoker of big cigars and wearer of brown derbies and suits with wide stripes was in fact the best hope that existed for the actual enactment into law of the social welfare measures for which they had fought so long with so little real success." The Catholic son of immigrants, Smith was raised in abject poverty on Manhattan's Lower East Side and was educated, he said proudly, at the Fulton Fish Market, not high school. By 1904 his "gift o' the gab" in public speaking and the heft of Tammany influence had elevated him to the State Assembly.

Smith was still Tammany's lackey in 1910 when Governor Hughes declared that the state should build power-generating reservoirs on streams in state parks, and gradually Smith came to believe, with Hughes, that the state's undeveloped waterpower "should be preserved and held for the benefit of the people." To that end, Hughes had secured the passage of legislation "authorizing and directing the state water supply commission to devise plans for the progressive development of the water powers of the state for the public use and state ownership and control." A preliminary plan was proposed to open up a navigable seaway along the St. Lawrence River that was compatible with power generation. The plan was rejected by the legislature, ostensibly because of its cost, though much of that political resistance was owed to Wall Street and Tammany influence. At the same time, for the next *forty years* the railroad interests fought against the seaway to protect their profits from rail cargo between the Midwest and the East Coast ports and markets.

Smith began to identify more solidly with Progressive causes after leading an investigation into the 1911 fire at the Triangle Shirtwaist Factory that killed 146 people. Workplace safety led him to support other Progressive causes like women's suffrage, addressing social needs, and more efficient government. Gradually Smith was joining the Progressives of the era, who included Republicans such as TR and Charles Evans Hughes and such Democrats as Franklin Roosevelt and Woodrow Wilson.

When Theodore Roosevelt formed the Progressive "Bull Moose" Party to run in 1912, he continued to press for the public ownership of water power, a stance he maintained even after Woodrow Wilson won the election. Two years later, TR spoke in upstate New York: "You have in this section a most valuable asset in your natural waterpower.... You have elected too many men in the past who have taken what belongs to the nation. Coal and oil barons cannot compare to waterpower barons. Do not let them get a monopoly on what belongs to this state. There has been a persistent effort to give private corporations control of the water power in this country."

That endorsement of public power by TR was endlessly repeated by Al Smith throughout his four two-year terms as governor. From his first day in office he pressed the state to harness the potential of the Niagara and St. Lawrence rivers and other inland waterways, insisting that "the state must itself retain ownership and control of waterpower at its source if the people and not private interests are to be the real beneficiaries by its development." Smith's waterpower policy was rejected by the state legislature in 1919 and 1920 and (after one term out of office) rebuffed again in 1923.

Nonetheless, in his annual message on Jan. 2, 1924, Smith proposed the creation of a State Water Power Authority "to take over and develop the state's power resources in accordance with a plan to be submitted to the Legislature for approval by law." The Water Power Authority would finance itself by issuing bonds to private investors and could never make use of taxpayer dollars or state credit. It was fundamentally different from other government agencies in that the state could not be held financially liable for any debt the Authority might incur.

An "Authority" was a new kind of hybrid: a government-controlled entity that serves the public while functioning as a self-sufficient corporation. England's Port of London Authority, created in 1908, was the first such public trust to be chartered by the Crown, which gave this corporation the *authority* to operate. In 1921, the Port of New York Authority was established along similar lines, as a "public benefit corporation" tasked with maintaining the public infrastructure and bureaucracy of New York and New Jersey transportation, much like a municipal corporation. So as not to endanger Edison and other private utilities, the Water Power Authority that Smith envisaged would *not* distribute power directly to the consumer, but it would oversee, own, and control the development of power at its source.

The State Senate approved the necessary legislation, but the Assembly, where Wall Street and its Tammany cohorts ruled, refused to pass it. Smith repeated his efforts for the next four years in a row and was defeated every time. Even in his final annual message to the state legislature in 1928, Smith, dubbed "the Happy Warrior," did not relent. "Giant power combines naturally will stand against the proposal of a public authority. On the other hand, there is today an insistent and growing demand for the development of these power resources by their rightful owners—the people themselves."

In 1928, when Al Smith left the Governor's mansion in his bid for the White House, he turned to Franklin Roosevelt. Smith believed that New York's gubernatorial election would determine how the state voted for the Presidency, and he could hardly afford to lose his home base, knowing that across the country many would reject a Catholic candidate out of hand. FDR had shown loyal support for Smith in his last two elections for governor and had delivered the nominating speech on his behalf at the 1928 Democratic National Convention. But his struggle with polio remained FDR's primary consideration, and it was only one month before the convention, under enormous political pressure, that he allowed his name to be put up. He was nominated by acclamation at the party's convention, and campaigned hard across the state through the autumn. That November, FDR was elected governor by just 25,000 votes out of four million cast. Not only was Al Smith defeated in the national election, he also lost in New York.

FDR Takes Over

Beginning with his inauguration, Governor Roosevelt seized the baton that Governor Hughes had passed to Governor Smith, driving home his conviction "that the water power of the state should belong to all the people. No commission, no, not the Legislature itself, has any right to give ... a single potential kilowatt in virtual perpetuity to any person or corporation." And he insisted that it was "the duty of our legislative bodies to see that this power, which belongs to all the people, is transformed into usable electrical energy and distributed to them at the lowest possible cost."

It was, said FDR, "intolerable that the utilization of this stupendous heritage should be longer delayed by petty squabbles and partisan dispute," and he wanted to make sure that New Yorkers—as consumers and voters—understood exactly what those disputes were about:

> There remains the technical question as to which of several methods will bring this power to our doors with the least expense. Let me here make clear the three divisions of this technical side of the question. First, the construction of the dams, the erection of the powerhouses and the installation of the turbines necessary to convert the force of the falling water into electricity; Second, the construction of many thousands of miles of transmission lines to bring the current so produced to the smaller distributing centers throughout the state; and Third, the final distribution of this power in thousands of homes and factories.

> How much of this shall be undertaken by the state, how much of this carried out by properly regulated private enterprises, how much of this by some combination of the two, is the practical question that we have before us. And in the consideration of the question I want to warn the people of this state against too hasty assumption that mere regulation of public service commissions is, in itself, a sure guarantee of protection of the interest of the consumer.

That FDR chose to fill his first inaugural address with such specific terminology demonstrates the enormous importance he gave to the matter in 1929. And he was acknowledging that, in the decades since Governor Hughes had created the Public Service Commission, it had had only limited success reining in the commanding political and economic influence of the holding companies, which had grown exponentially into an electric power empire. While the states regulated the operating companies, no one regulated the holding companies, which the great comedian of the era, Will Rogers, described as "a thing where you hand an accomplice the goods while the policeman searches you." By 1932, according to energy analyst Leonard S. Hyman, "the eight largest holding companies controlled 73 percent of the investor-owned electric business."

"At the top of the industry," wrote historians Richard Rudolph and Scott Ridley, "were the equipment manufacturers, led by General Electric, which controlled the stock of hundreds of local power companies. Also tied in at the top levels were the major banks and investment houses that supplied the money for constant expansion, and the fuel companies that provided coal and later oil and uranium for the power use that doubled every decade." Local banks also profited from granting large loans to power companies, which enhanced their local political influence.

The private utilities collaborated through the National Electric Lighting Association (NELA) to fund lobbying, elect local candidates to derail regulatory action, and prevent public control of their monopolies. In 1926 alone, NELA sent out speakers to address 18,423 audiences at

Rotary, Kiwanis, and Lions clubs, reaching 2.5 million people face-to-face, with one consistent message: private power was the only reliable route to the future.

Nationally, Rudolph and Ridley noted, "the empire's influence ran from local city halls to the corridors of the federal government. Each power company was given a monopoly right to a geographic territory by a city or town. To protect those rights and expand control of power resources, such as the nation's rivers, the empire developed a system of deepening economic and political influence." Any attempts to establish public nonprofit systems "were steadily undermined by the private power monopolies and their corporate allies." Pennsylvania Governor Gifford Pinchot insisted that "nothing like this gigantic monopoly had ever appeared in the history of the world," and it seemed unstoppable.

Two months after his inauguration as governor, FDR proposed a commission to study development of waterpower; specifically, he said, "it seems to me best at this time to focus recommendations and public attention on the development of the St. Lawrence River.... The great preponderance of public opinion supported immediate action," he insisted, and this commission, to be called the Trustees of the Water Power Resources on the St. Lawrence River, "should be composed of men in whom there is great public confidence," including former governors Hughes and Smith. The trustees would report their findings ten months later and, with legislative approval, would carry out their plan. By the standards of the era, this was a "very brave step in the right direction," averred Nebraska Senator George W. Norris, the leading champion of public power in Congress. But FDR faced Republican majorities in both the State Senate and the Assembly, and the political opposition fostered by the electric power empire was waiting with sharpened knives. The bill to form such a commission died in committee and never made it to the Legislature.

FDR responded by going over the heads of the legislators to the voters themselves; he was learning how to do this better than any politician in US history. "Throughout the summer [1929] Roosevelt toured the state," wrote biographer Ernest K. Lindley two years later, "making from four to eight speeches a day, visiting towns and villages and crossroads which had not seen a Governor in years, making himself known personally to voters of both parties, preaching 'cheap electricity.'" And three years before his first "fireside chats" from the White House, he was practicing his radio technique from the Governor's Mansion. FDR's personal outreach and out-of-season campaigning was unrelenting. Eschewing any highfalutin' political philosophy, he repeatedly hammered home one simple fact: New Yorkers were paying on average *eight times* as much as Toronto's public power customers. That was a message not even Republican voters could ignore.

Then, in October, the stock market crashed. Margin calls on all the leveraged stock triggered panic, and within two weeks the Dow-Jones Index was reduced by 40 percent and falling. Jobs evaporated as businesses collapsed and the entire overextended economy began to shrivel.

As Warren Buffett famously observed many years later, "Only when the tide goes out do you discover who was swimming naked." With the near-death experience of Wall Street, it became obvious that political influence had allowed the power monopoly to operate much the way Charles Ponzi had in 1920: pocketing the investors' money and repaying them excellent interest from later investments. But with the run on the banks any new investors vanished, and by the early 1930s a leading figure in the power empire had become a fugitive from the law and an icon of power industry corruption.

Samuel Insull had emigrated from England to work as Thomas Edison's secretary even before the first central plant was built on Pearl Street. But in 1892, when J. P. Morgan and Vanderbilt offered inducements to abandon his mentor, Insull quickly accepted and moved to Chicago, where he was made president of Chicago Edison. There, Insull instituted a crucial principle in power generation imported from England: Diversity of load creates greater savings and profits. In most cities, for example, the electrical demand for traction (trollies and trams) peaked during morning and evening rush hours, while elevators and other motors operated mostly through the day, and lighting predominated after dark. So the peak load was never the sum of all three demands, just (approximately) the highest of the three. With that understanding, and with price inducements to customers, Chicago Edison could realize significant savings in generation. Unlike J. P. Morgan, who would have wanted to sell a separate generator to every business and household, Insull preferred the economies of scale achieved by large central plants.

Insull also had a surprising attitude toward regulation: he welcomed it. As early as 1898 he spoke at an industry convention and promulgated the view that competition was "economically wrong" for power companies, because generation and transmission are a "natural monopoly." His argument was that a single utility could serve each community more efficiently than competing companies that each purchased expensive generators, hung power lines, and dug up the streets. As well, the customers of a regional monopoly would benefit from the economies of scale that only a large, single producer could achieve, because, as the scale of production increases, unit costs decrease; that is, *if* the producer passed the savings along. Still, Insull insisted, "while it is not supposed to be popular to speak of exclusive franchises, it should be recognized that the best service at the lowest possible price can *only* be obtained ... by exclusive control of a given territory being placed in the hands of one undertaking." His colleagues listened in stunned amazement.

Insull had recognized that "it would be necessary to give the state regulatory agency the right to limit their profits," wrote Rudolph and Ridley, and that accepting that agency would "head off the growth of publicly owned systems by offering an alternative form of public control over the electric business." Although he did not mention it at the conference, Insull was confident that he could control any regulatory body, because his influence in the state and federal legislatures often allowed him to recommend or even select the most pliable commissioners to oversee his utilities. How he bought that influence was no secret: Insull freely acknowledged to a US Senate committee that he had contributed $125,000 to the Senate campaign of the chairman of the Illinois Commerce Commission, which regulated his businesses. There was no law forbidding it, and similar donations were common in other states.

Insull had always known that the real profits were in the holding companies, which, by this time, were competing to buy operational plants and other businesses. Each time his holding company purchased a business, he inflated its value tenfold, and then sold stock in it. As Hyman explains, "A good promoter could put together a combination of properties, sell preferred stocks and bonds to the public to pay for the properties, take ten percent or more as commission, and keep the bulk of all of the voting common stock of the holding company, thereby remaining in control without having paid a cent into the business.... By bloating the value of the power companies in questionable stock deals, bankers and stock firms... made huge profits that could be passed on as higher prices to consumers," wrote Rudolph and Ridley, while "the favor and support of industrialists was gained by granting them cheap rates." That naturally added to the exorbitant rates borne by the average customer. As the US ambassador to Germany exclaimed at an international energy conference: "I know of no other manufacturing industry where the sale price to the great mass of consumers is 15 times the cost of production of the article sold."

Insull's personal empire, wrote Hyman, "operated in 32 states, and owned electric companies, textile mills, ice houses, a paper mill and a hotel. With a capital investment of only $27 million, Insull controlled at least $500 million of assets in 1930." That degree of leverage, largely in junk bonds, was common on Wall Street before the market crashed; indeed, that is exactly *why* it crashed. As the value of the magnate's utility stocks fell, so did the value of all his collateral. He managed to keep borrowing from "friendly" New York banks against the shrinking value of his shares, without knowing that those banks, which also faced liquidity problems, were acting on behalf of Morgan. When Insull's Ponzi scheme fell apart, some 600,000 of his investors lost all their savings. In short order J. P. Morgan took control of Insull's companies just as he had Edison's, Westinghouse's, and Tesla's. In 1932, Insull fled to France.

FDR was soon using Insull as the poster child for the private utilities' greed. One year after his first inaugural address in Albany, he repeated his plea that St. Lawrence waterpower "remain forever in the actual possession of the people of the state or of an agency created by them ... to insure a fair and reasonable rate to the consumer, especially the household users." But "sensing that he held the whip hand," according to biographer Lindley, he extended his reach beyond power generators to state-owned transmission lines as well, declaring that he "thought it advisable that the state agency should at least provide the financing of and retain the fee to any system of state-wide transmission of electricity made necessary by the new power development."

There was a new urgency to his effort because, Lindley explained, he had known for some time that "during the summer the consolidation of the upstate power groups into Niagara Hudson Power Corporation had been completed under the guiding hands of the House of Morgan and the corporation had acquired the riparian rights to the St. Lawrence site." Riparian rights entitle landowners along the river banks to make reasonable use of that water. Still, Morgan would need an agreement with Canada that would involve the Federal government, not to mention state licensing. A power development would also require a vast capital investment that neither Morgan nor the state of New York could expect to raise in 1930. Perhaps that is why "the bankers intimated that, given a 'sound' plan, they would not be averse to public development of the St. Lawrence." Republicans were still the majority in both state houses, but their attitude had been tempered by the ongoing economic collapse and the cumulative impact of FDR's campaign for public power. Rural Republicans had heard the Governor's message loud and clear and were demanding cheap electricity from their legislators, some of whom had come to accept the idea of public power. Fearful of possible losses in the November election, they knew they could no longer ignore the issue of public waterpower: capitulation was their only option. However, rather than reintroduce FDR's plan, on Jan. 13, 1930 the Republicans proposed their own bill for a commission to "devise and report a plan or plans for the development of hydroelectric power on the St. Lawrence River."

"This is one of the happiest days of my life," FDR exulted, "and one of the most important for the people of the state of New York." He promptly sent a telegram to Al Smith, asserting that the Republican bill reflected "the great basic principle for which you and I have fought so long. There is no doubt it is a great victory." Two months later the Legislature approved the bill creating the St. Lawrence Power Development Commission, which FDR "adjusted" to his liking and signed into law on March 29. "It is a milestone," he said during the signing, "marking the end of a 20-year struggle against great odds, for it takes the first step towards securing cheaper electric light and power." New York Secretary of State and Roosevelt confidant Edward J. Flynn certified the act establishing the Commission. He could not have imagined that more than a half a century later his son, Richard Flynn, would become chairman of the New York Power Authority. The Commission also included labor attorney and activist Frank P. Walsh, who would soon become the first chairman of the Power Authority. The commission's legal advisers included a young Charles Poletti, who, years later, would become the only person to serve both as New York governor and as a Power Authority trustee.

The act specified unambiguously that the five commissioners were to be chosen by the governor, not the legislature; without this key provision, the commission would have been stacked with Republicans and the Power Authority might never have gone any further. FDR judiciously appointed two Republicans (who he knew supported public power), and three Democrats, including Walsh. In August 1930, FDR joined all five at the proposed hydroelectric site on the St. Lawrence River.

While awaiting the commissioners' report endorsing the power project, which was almost a foregone conclusion, FDR undertook the revival of the Public Service Commission, which had devolved into a zombie agency during the nine-year tenure of its Republican chairman, William A. Prendergast, who finally felt compelled to resign. In a speech the next day, FDR disparaged Prendergast in no uncertain terms:

> [He] said that it was a function of the Public Service Commission to sit up on a bench and hand out justice on the one hand to

> the people of the state, and on the other side to the utilities ... a sort of arbiter between two contesting forces. Historically, practically, legally and in every other way, Mr. Prendergast was dead wrong.... The Public Service Commission is the representative of the Legislature, and back of the Legislature, of the people. It is not dealing between two contestants.... [its] sole function, as the representative of the people of this state, [is] to see to it that the utilities do two things: First, give service, and secondly, charge a reasonable rate. We are going to get some teeth in the Public Service Commission Law.

Naturally, much of FDR's time during his first two-year term as governor was focused upon the impact of Wall Street's implosion and the Depression that swiftly followed. Unemployment relief was one of many contentious issues he pressed upon the Republicans and generally prevailed, thanks to his solid Democratic support that included Tammany, his old foes as state senator. As biographer Lindley put it drily, "The fad of his youth, attacking the machine, had given way to the more profitable practice of using them." Summoning all his political clout, FDR passed the first state unemployment relief, launching the Temporary Emergency Relief Administration (TERA) with $20 million, and soon after he appointed Harry Hopkins to lead the agency.

In his successful campaign for a second two-year term as Governor in 1930, FDR again traveled the state, emphasizing the urgent need for public power, pointing out again and again that New Yorkers paid an average $25.63 a month while Toronto residents paid $3.40 for the same kW-hours, and noted the "ever-growing, ever-insistent public opinion ... demanding its right and due in the form of cheaper electricity." In Binghamton, he promised that housewives "will have the benefit of electric lights, and of an electric refrigerator, an electric range, electric vacuum cleaner, electric radio, dishwasher, clothes washer," and pointed out that the proposed waterpower project on the St. Lawrence River could provide enough electricity for half the state.

The St. Lawrence River Undertaking

As they prepared their report, the St. Lawrence Power Development commissioners met with President Herbert Hoover, as did officials of the Federal Power Commission and the Hydro-Electric Power Commission of Ontario. In 1930 Hoover was already deeply embroiled in the issue of public power: Just then a consortium of contractors hired by the US government was beginning construction of Boulder Dam, later renamed Hoover Dam. Approved by Congress and President Calvin Coolidge in 1928 after years of planning, the dam along the Arizona-Nevada border was the largest concrete structure ever undertaken, creating Lake Mead, the largest reservoir in the United States. This was indisputably a federal project because the Colorado River bordered on seven states that would be significantly affected, in a region where water is scarce. Of course, as the Depression deepened, the projected cost of the dam—$49 million—was growing more onerous.

But the most complex hydropower issue on Hoover's desk was Muscle Shoals in the Tennessee Valley. This had become the central arena in the battle pitting public power against the monopoly-owned utilities, in which Senator Norris focused upon the utilities' high rates against a background of rural, Depression-era poverty for which hydropower, and even the short-term construction jobs it would create, appeared to be a life-or-death issue for whole communities. Norris "exhibited photostat copies of two domestic, monthly bills stamped 'paid,' each for 334 kW-hours of electricity," wrote Judson King, an important FDR consultant on the Rural Electrification Administration and Tennessee Valley Authority. "The Canadian bill was for $3.55 while the US bill was for $23.18." Norris repeatedly cited such examples of the price difference between public power and private power for years.

When TR's Inland Waterways Commission Report had asserted federal control over navigable waters in 1908, it had left regulation of hydropower to the individual states, which usually allowed private industries to build and operate dams "for nothing and forever." During World War I, President

Wilson had invoked his wartime authority under the National Defense Act and the Army began construction of Wilson Dam at Muscle Shoals to produce explosive nitrates, which required a great deal of electrical power. The dam was the first of nine that would later make up the Tennessee Valley Authority's hydropower system. It extended across half the shoals, but only began producing power years later from three of its planned eighteen generators. After 1920, federal licenses were required for all dams from the new Federal Power Commission (which later became the Federal Energy Regulatory Commission). Meanwhile, opponents of public power used the US Senate as their means to block any further construction.

In the next decade the debate over government-owned power rose to a new pitch of acrimony. Two Congresses had passed bills demanding that the federal government should remain in control of the Muscle Shoals project and operate its power facilities. Two presidents, Coolidge (who took office when Harding died in 1923) and Hoover, had vetoed those bills, the latter insisting that government-owned industries were "socialistic," a potent condemnation in the years after the Bolshevik Revolution in Russia.

As for the St. Lawrence project, Hoover's attitude was more complicated. Like Coolidge and Harding before him, he ardently supported development of a St. Lawrence Seaway, hoping that would win him votes from disaffected farmers in the Midwest who sought an alternative to the railroads' high prices for shipping grain eastward. At the very least, Hoover would appear to make the effort without actually jeopardizing his support from the rail magnates. And while it obviously made economic sense to construct a seaway and a hydropower plant at the same time, few politicians supported both. In general, Republicans were reluctant to support Democratic solutions and vice-versa. Hoover was especially wary of Governor Roosevelt's plans for St. Lawrence hydropower and adamantly opposed state ownership of electric distribution. FDR's progressive politics were abhorrent to his own damn-the-torpedoes defense of free-market economy, even as the Depression deepened. Also, Hoover knew that Governor Roosevelt was hoping to terminate his own presidency in the 1932 election.

Hoover's views aside, the St. Lawrence endeavor, comprising both the seaway and the power project, was complicated by the involvement of a foreign country, Canada, and so required assent from the US Senate, where a broad coalition of private power and railroad lobbyists could block the undertaking. The seaway would not only cut into the profits of railways and the east coast ports, but also those of Mississippi barges to the Gulf of Mexico and intermediary hubs like Buffalo, New York. Opposition also arose from related labor unions and isolationists, who did not want the seaway to open the heart of the country to foreign vessels. In 1930, all those lobbying interests formed the National St. Lawrence Project Conference, an umbrella organization to oppose the seaway.

Canadian opposition to the project mirrored that in the United States, wrote Daniel Macfarlane in *Negotiating a River: Canada, the US, and the Creation of the St. Lawrence Seaway* (2014). When Liberal party leader Mackenzie King became Prime Minister in 1921 for the first of his twenty-two years in office, "he shot this project down without even the courteous pretense of careful consideration," expressing concern that it could cause "domestic discord." With a Canadian government unwilling to even discuss the seaway or hydropower during the years that followed, there was nothing that Hoover could do.

Signing the Power Authority Act

As expected, the report issued by the St. Lawrence Power Development Commission in January 1931 called for the establishment of a Power Authority to build power plants beside the river. They had, wrote Lindley, "a new engineering plan permitting construction of the dam on dry land and the subsequent diversion of the river," which could save $70 million in construction costs. FDR promptly hailed the report as "a great step forward" and declared "the time is ripe for the creation by the Legislature of a Power Authority."

The report specified that transmission and distribution of power were to be provided by private utilities, which would placate Hoover.

But the minority report from Commissioner Thomas F. Conway "urged that the Power Authority be endowed with the right to build and operate transmission lines in the event that suitable terms could not be obtained from the utility companies." He also proposed that "ten percent of the power be reserved for the publicly owned distribution systems of municipalities and other local subdivisions. On these points, FDR stood with Conway."

He would not budge, FDR explained, because the possibility of building state-owned transmission was "the only club in [the Power Authority's] possession in its negotiations with the present utility monopoly. If it did not have these alternatives, the state would be at the complete mercy of the Niagara Hudson Power Company.... I believe these two alternatives provide the whip hand—the trump card." This project would represent a collaboration between government and private enterprise, and would "remove the property from the field of speculative profits...."

Six weeks later, it was a Republican from beside Lake Ontario who introduced the Power Authority bill in the Assembly, where it passed quickly. The State Senate also approved the bill, but there Republicans added a "poison pill" amendment empowering the Legislature, not the Governor, to appoint the Power Authority's five trustees; of course, those appointees could delay the project indefinitely. FDR was given only two options: either approve the amended bill or veto it and postpone any progress for another year, as the Republicans hoped.

Once again, he had another plan: to appeal directly to the people to save the proposed agency. "Within twenty-four hours the Republican leaders were submerged in telegraphic and editorial protests from civic organizations, chambers of commerce and individuals, especially in their own territory in northern New York." FDR's threat to give a radio speech to "lay the facts before the people" prompted Republicans to strike the amendment just hours before the broadcast.

On the radio, Roosevelt hailed this as a victory over the monopolies. "Stronger than all these put together is the influence of Mr. and Mrs. Average Voter," he said. "It may take a good many years to translate this influence of the people of the state into terms of law, but public opinion, when it understands a policy and supports it, is bound to win in the long run." The Senate approved the bill unanimously in exactly the form that FDR insisted upon, and on April 27, 1931, he signed the Power Authority Act into law at his home in Hyde Park. In a radio address that week he declared, "I place first of all in importance the passage of the bill creating the Power Authority to develop the state-owned waterpower on the St. Lawrence River.... It is my earnest hope that this is the forerunner of cheaper electricity for the homes and farms and small business people of the state."

"There is hereby created a corporate municipal instrumentality of the state," the original Act declares. The unusual word "instrumentality," which sounds evasive, distinguishes this organization from a government agency. Instead, an instrumentality (like the Federal Reserve Banks, Fannie Mae, or the Wilson Center) is a corporate organization devoted to a public purpose, and while it is chartered by state or federal government, it does not receive taxpayer dollars; indeed, "the object and purpose of this statute is that the said project should be in all respects self-supporting."

The Authority was not entitled "to pledge the credit of the state, nor shall any of its obligations or securities be deemed to be obligations of the state." Instead, it must raise its funds from investors by the sale of bonds, which carry one significant benefit: "The securities and other obligations issued by the Power Authority, their transfer and the income therefrom shall, at all times, be free from taxation within this state." Those bonds, however, could be offered on the market only *after* contracts for the sale of power had been signed; and those contracts were to be the subject of public hearings to insure complete transparency. The Act also specifically endowed the Authority with the power of "eminent domain" to obtain lands that it needs "for a public use."

A Template for National Reform

FDR had at last succeeded in creating the legislative means to produce public power, almost a quarter century after Governor Charles Evans

Hughes had taken the first steps toward that goal. He had also learned a vast amount about the technology of power generation, and developed a practical and legislative model that would serve well for national energy programs also. The following year he made the matter of government-owned utilities a key issue in his presidential campaign, saving his most impassioned speech for Portland, Oregon, in September 1932. "State owned or federal owned power sites can and should and must properly be developed by government itself," he declared. "Private capital should, I believe, be given the first opportunity to transmit and distribute the power on the basis of the best service and lowest rates *to give a reasonable profit only*. The right of the federal government and state governments to ... transmit and distribute where reasonable and good service is refused by private capital, gives to government—in other words, the people—that ... essential 'birch rod' in the cupboard." Again and again he reminded his audiences of the perfidious Samuel Insull, holding him up as the perfect symbol of the power monopolies.

Throughout Hoover's campaign, he remained unwavering in his opposition to government-owned power and steadfastly defended his veto of the Muscle Shoals bill. But as the Depression cut deeper, that issue paled in significance beside the urgent and desperate plight of the nation. Polls showed that Republicans faced defeat in most every race. Hoover's use of radio was ineffective and owing to his years of inaction, on his campaign tours he was pelted with rotten eggs and fruit and faced more than one assassination attempt. In November, FDR won the presidential election with 57.4 percent of the vote to Hoover's 39.7 percent, 26 points lower than his percentage in 1928.

After the election, but even before he took office, FDR proposed the creation of the Tennessee Valley Authority as an instrumentality of the federal government. He even visited Muscle Shoals before his inauguration, knowing this would be ground zero in his battle for public power and the jobs and low-cost electricity it could produce. There was ample domestic opposition to the St. Lawrence project, but it could not be budged forward at all as long as the Canadian government refused to discuss it. And it was far from clear that sufficient capital could be persuaded to buy untested bonds in the depths of the Depression.

President Franklin Roosevelt was sworn into office on March 4, 1933, by the Chief Justice of the Supreme Court, none other than Charles Evans Hughes, whose court would prove to be the single greatest obstacle in FDR's presidency.

CHAPTER SIX

THE LONG HAUL, 1931–1954

After the Power Authority was created in 1931, twenty-seven years passed before any electricity was generated by the State of New York. A variety of factors contributed to the long delay, including the political influence of the private utilities, railroads, and affected labor unions, as well as the Depression and World War II. But a primary obstacle was the obduracy toward the seaway project from numerous interests in Canada.

The Great Lakes–St. Lawrence Deep Waterway Treaty

Mackenzie King was prime minister for all but five years (1930–35) from 1921 to 1948, an extraordinarily long tenure. His Liberal Party was often divided on the St. Lawrence issue, but King himself, though he wavered, generally opposed the project because of both its enormous cost and the strong resistance it engendered in Québec. Strictly speaking, as Canada was part of the British Empire, its foreign policy was the prerogative of the British Crown, but in reality London kept a light hand on the tiller and left such local decisions to the Parliament in Ottawa. There, ardent nationalists argued that the St. Lawrence should remain a "Canadian river," and that the project would in reality prove to be a "Trojan Horse" from the

United States. The *Montréal Gazette* insisted that a joint project with the United States would result in a loss of Canadian autonomy. "Our great river must be ours and ours alone," declared an Anglican canon in Québec City. Canadian railway interests also applied political pressure to prevent any agreement on the seaway. "The United States was the most consistent suitor," writes Canadian historian Daniel Macfarlane, "but was frequently met with Canadian ambivalence and obfuscation, particularly when King was Prime Minister."

But in 1930, during King's five-year "interregnum," the new Conservative prime minister, R. B. Bennett, supported the seaway, and unlike the Liberal Party, the Conservatives' majority in Parliament did not rely upon support from Québec province. Despite Bennett's personal distaste for President Hoover, the Conservative Party was willing to negotiate a treaty with the United States, in part, wrote Macfarlane, because of "the possibility that the United States would proceed with its own national waterway" and the growing likelihood that Hoover would not be reelected. While it was well known that FDR ardently supported St. Lawrence hydropower, his attitude toward the seaway, and that of his supporters, seemed uncertain.

So Bennett and Hoover pressed ahead, and on July 18, 1932, just months before FDR won the presidential election, the Great Lakes–St. Lawrence Deep Waterway Treaty was signed by the two countries, pending ratification by the US Senate, as constitutionally required, and the Canadian Parliament. This enormous engineering project centered upon the waterway, twenty-seven feet deep, from the Great Lakes to Montréal; locks, canals, and dams were specified and the division of labor between the two countries was detailed. The greatest challenge lay in the International Rapids section, where there are sudden breaks in the gradient. The Treaty prescribed the construction, in two stages, of powerhouses on both sides of the international border that runs down the center of the river, generating a stupendous 2,200,000 horsepower, to be equally divided. However, as Macfarlane notes, "the treaty deliberately left open-ended the precise manner in which the power was to be distributed, given the jurisdictional federal-state and federal-provincial issues in each country." Ratification by both governments was likely to hinge upon that ambiguity.

No less important would be the approval of the International Joint Commission (IJC), an independent committee created in 1909 when the United States and Great Britain signed the Boundary Waters Treaty, which rules upon and regulates projects affecting boundary or transboundary waters and may regulate the operation of these projects with a special concern for preserving environmental quality. Both Canada and the United States appoint three of the six IJC Commissioners, who, after being chosen, operate independently and at arm's length from their respective national governments, working by consensus toward solutions in the best interest of both countries.

As well, the US Federal Power Commission and its Canadian counterpart would also have to allow the construction, which was far from certain. There were also serious questions about special powers granted to the St. Lawrence Commission that would be created if the Treaty took effect, and uncertainty about what legal jurisdiction would pertain should either country fail to fulfill its obligations for any reason.

Still, the most contentious issue was the cost: $543 million, a monumental sum in the damaged economy of 1932. The two countries would share the cost "almost equally," but with one major proviso: Canada would be credited with work already performed, raising the US obligation to the lion's share of more than $400 million. Clearly that was going to be an Achilles' heel in seeking ratification, especially for the dozens of states that would not benefit at all from this expensive undertaking.

The Treaty was still pending when FDR took office in March 1933, by which time fifteen million Americans were unemployed. His first one hundred days famously began with an extraordinary flurry of presidential actions, starting with the bank closings to stop the panicked run on them. On his thirty-seventh day, FDR called upon Congress to create the Tennessee Valley Authority, "a corporation clothed with the power of government but possessed with the flexibility and initiative of a private enterprise." Still, he made no mention of the St. Lawrence project, nor did he submit the signed Treaty to the US Senate for ratification for almost a year.

The Canadian Parliament postponed voting on ratification until after the Senate had.

Notwithstanding FDR's apparent indifference to the seaway project, the Treaty advanced, if slowly, and in March 1934 the US Senate approved the Great Lakes–St. Lawrence Deep Waterway Treaty by 46 votes to 42. That, however, was insufficient since treaties require a two-thirds majority (at that time, 64 votes) for ratification, and the bill went down in defeat. In retrospect, it is clear that FDR felt he needed to apply all his political capital for public power to the TVA project, which would have a quicker, greater economic impact upon a region deep in despair with 30 percent unemployment.

It is also possible that the St. Lawrence project as planned in the early 1930s was technologically ahead of its time. Without the equipment, materials and techniques that were available by the 1950s, time and budget overruns might have proven disastrous. Of course, we will never know.

The Tennessee Valley Authority

FDR saw much more potential in the TVA than in electricity. Judson King, who helped establish and manage this Authority legally, politically, and logistically, described the President's broad vision: "His unique contribution, later amplified, was regional planning to conserve natural and human resources in one comprehensive enterprise." In fact, apart from navigation, flood control and hydropower, the TVA project would serve as an economic development agency throughout most of Tennessee and parts of Alabama, Mississippi, Kentucky, Georgia, North Carolina, and Virginia, a region with nine million inhabitants today, largely thanks to the TVA. One of the region's urgent needs for electricity was in the production of nitrates from atmospheric nitrogen to produce fertilizer for the whole region, where the soil was mostly depleted by more than a century of intensive cotton and tobacco production. But beyond that, said FDR, the project, "if envisioned in its entirety, transcends mere power development.... It touches and gives life to all forms of human concerns." All this would be achieved without taxpayer dollars, through the sale of bonds and low-cost electricity. The TVA Act specified that the government did not guarantee any debt it might incur, though, as the *Economist* put it recently, "creditors and ratings agencies have never really believed that Uncle Sam would let Uncle Dam default."

Senator Norris advanced the bill through the Senate Committee on Agriculture, while Congressman John J. McSwain of South Carolina introduced the bill in the House Military Affairs Committee, before which then-Democrat Wendell L. Willkie appeared to express his ardent opposition. Willkie was president of a New York holding company that operated many power plants around Tennessee, and for the next seven years he coordinated and spoke for the region's investor-owned utilities. Of course, Willkie, who supported FDR at the Democratic National Convention in 1932, is best remembered for becoming his Republican opponent in the 1940 election.

Several southern power companies like Willkie's actually supported TVA development, wrote King, "provided the government did not build transmission lines to market its power." This became the key political battle, as it had been with New York's Power Authority Act. Once again FDR, goaded sharply by Senator Norris, insisted it was crucial that public power have the legal right to transmit power or it would be held captive by the investor-owned utilities. And once again FDR had his way: on May 18, 1933, he signed the Tennessee Valley Authority Act into law. With that, wrote Judson King, "the 35-year legislative warfare over who should control development of the Tennessee watershed had ended—for the time being."

FDR passed many new laws in his first hundred days, but the TVA Act, according to historian Arthur M. Schlesinger, Jr., "expressed more passionately a central presidential concern, [which] arose only in part from Roosevelt's old absorption with land, forests, and water. It arose equally from his continued search for a better design for national living.... As Governor of New York, he had talked of redressing the population balance between city and countryside—taking industry from crowded urban centers to airy villages and giving scrawny kids from the slums opportunity for sun and growth in the country. The Depression and the presidency provided new opportunity to move toward a 'balanced civilization.'"

FDR's Power Projects

The TVA project benefited enormously from the lessons FDR had learned as governor while developing the Power Authority and planning the St. Lawrence project. The same could be said of the Grand Coulee Dam. By July of 1933, as construction proceeded on Boulder Dam (later renamed Hoover Dam) along the Colorado River, the same consortium of contractors began work on the first dam at Grand Coulee on the Columbia River, and in September plans were approved for the Bonneville Dam, forty miles east of Portland, Oregon. When construction began the following year, 3,000 men listed on the rolls of unemployment relief were hired at fifty cents per hour. But even as construction advanced, the region's investor-owned power industry waged war over transmission and distribution until 1937, when the Bonneville Power Administration was established to market the output of the federal dams on the Columbia and the region's other rivers. The BPA was initially administered by J. D. Ross, a public power pioneer who had served as a consultant for the New York Power Authority.

Meanwhile, despite intermittent efforts at revising the St. Lawrence Waterway Treaty to improve its prospects in the US Senate, no substantial progress was made. When Prime Minister King returned to office in 1935, President Roosevelt succeeded in cultivating a friendship with him; or as historian Macfarlane wrote, "FDR was a relentless flatterer and charmer, and the Canadian leader lapped it up." The two reached a major trade agreement, and they discussed modifying and broadening the unratified 1932 treaty to include diverting water for power at Niagara. But just as King became amenable, the premiers of Ontario and Québec were entangled in disputes over water rights, power contracts, and a variety of federal-provincial issues that made further negotiations pointless. In 1938, FDR suggested that the United States would carry out *all* the construction for the seaway and the US power project, leaving Canada with only the cost of building its own powerhouse; but when the Canadian government received that generous offer, it did not even bother to reply.

In the meantime, FDR marshaled all his political muscle to regulate the power monopolies' most profitable gimmick, the holding company. The Public Utility Holding Company (PUHC) Act of 1935 (which began as the Wheeler-Rayburn Bill) aimed to reform legal abuses by both the power companies and their Wall Street investors. The Act, wrote power historian Hyman, was intended to curtail "all the abuses [by the private utilities]: control of an entire system by means of a small investment at the top of a pyramid of companies, sale of services to subsidiaries at excessive prices ... intra-system loans at unfair terms, and the wild bidding war to buy operating companies." The Act also gave explicit, detailed regulatory powers over the holding companies to the Securities and Exchange Commission.

"The fight over [this] legislation they introduced," wrote Rudolph and Ridley, "would be the most fiercely fought battle of the New Deal, and it would touch off one of the most intense and scandalous lobbying campaigns in American history." Opposition to the bill was coordinated by the Edison Electric Institute (EEI, an industry lobbying group that still exists today), which instructed many thousands of investment bankers and insurance agents to write to their representatives. Freshman Missouri Congressmen Harry Truman received 30,000 letters demanding that he oppose the PUHC Act. Despite this unprecedented and well-funded political pressure, the Act passed both houses and, notwithstanding some compromises, it effectively reined in the monopolistic thuggery of the investor-owned utilities. The impact was undeniable: over the course of fifteen years it reduced the average national price of electricity from seven cents per kWh to less than three cents.

In 1935, the year in which the PUHC Act became law, FDR produced another of his greatest and best-remembered legislative accomplishments in the field of energy: the Rural Electrification Act, which granted federal loans for electrical distribution in the vast rural area of the United States. Perhaps nothing illustrated the problematic nature of private utilities better than the plight of the nation's rural population, especially on working farms and ranches, from the Atlantic Coast to the Pacific Ocean. Of course, it was true that planting poles and stringing wire forty miles across the countryside at triple the voltage (6900 volts) would increase the cost of providing service; what was untrue was the private utilities' argument that it was not economically feasible, and the Rural Electrification Administration proved it. The REA funded the distribution through new, member-owned cooperatives resembling municipally-owned utilities. The new agency was headed by Morris Llewellyn Cooke, one of the original New York Power Authority trustees.

The impact on the countryside, and thus on the country, was spectacular: in 1930 just 10 percent of US farms had electricity, but after a decade of the Rural Electrification Act, that number had risen to 45 percent. Teams of electricians spread out across the country and after bringing distribution lines to the farmhouses and barns, they installed indoor wiring as well, and mounted one lighting fixture in each room. This was an extraordinary feat by any measure: the federal government brought light into people's living rooms, bedrooms, and barns, transforming daily life and increasing productivity on farms and ranches across the country.

For all the profound changes in the domestic energy industry, the political impasse within Canada concerning the St. Lawrence development remained intractable. Then that, too, was finally resolved, not by the Prime Minister or the President, but by the Führer. After Germany invaded Poland in September 1939 and Canada, with Great Britain, declared war, the need for the seaway and increased generating capacity for the war effort was undeniable. When the inflexible premier of Québec was removed by election, the premier of Ontario suddenly became amenable, as long as his province could export some of its electricity. So, despite the re-election of Mackenzie King and his Liberal Party in 1940, Canada was ready to sign a new agreement that provided for a 27-foot-deep waterway from the head of the Great Lakes to the Atlantic Ocean.

But now FDR chose to postpone any action before his own reelection in November 1940 for an unprecedented third term. By the following May, after Ontario and Québec reached agreement, the St. Lawrence effort began again in earnest. After interprovincial agreement was achieved, the Power Authority "actively cooperated with the [US Army] Corps of Engineers in advancing this necessary preliminary work in concert with other public agencies of the Federal, Dominion, and Provincial Governments."

This time FDR sought to seal the deal with Canada as an executive agreement, which required only a simple majority in both houses of Congress and which he and King both signed. Significantly, FDR continued to see the power project and the seaway as indivisibly linked by the need for a dam, "the one simple fact to harp on," he said. "You cannot get power without a dam and you can't get a seaway without a dam. What is essential to defense, therefore, is the building of a dam."

The executive agreement was debated at length in committee and then reported favorably to the full House, where it was still waiting when Japan attacked Pearl Harbor in December. With that, FDR began promoting the project as part of the war effort, instructing defense agencies to "plan for full utilization of the St. Lawrence Seaway and Power Project both in our immediate and long-term defense efforts." But he was advised that first power could not be produced before September 1945 and defense projects were rejected unless they could be completed in 1943, which "knocks everything into a cocked hat," remarked Roosevelt with disappointment. The bill died in the House.

Nevertheless, the 1941 executive agreement would provide guidelines for the eventual development of the St. Lawrence, and parts of it were immediately put into effect by Canada: in some places diversions from the river were dug, including one at Long Lac, to deliver more water and produce more power at Ontario's new Niagara hydro plant, completed in 1939. But on the US side, preoccupation with the war effort understandably drained the energy from other efforts. Even New York supporters began losing hope after 1940, when the Legislature cut NYPA's annual budget to $50,000, leaving the Power Authority with a staff of three.

The Power Play for the St. Lawrence Seaway

With the death of FDR and the end of the war in 1945, his successor, Harry Truman, proved to be a staunch proponent for the St. Lawrence project, as was the Governor of New York from 1943 to 1954, Thomas E. Dewey. Of course, in 1948 Dewey tried but failed to replace Truman in the White House, but at least with regard to the St. Lawrence project, the two men saw eye to eye. Dewey was also in tune with Ontario's Premier, George Drew. Since 1945 the growth in demand for power on both sides of the border had been enormous; in Southern Ontario it had increased by 18 percent in just four years. So together, Dewey and Drew put forward a "power priority" plan that would allow the Hydro-Electric Power Commission of Ontario (HEPCO) and New York's Power Authority to develop power plants in the International Rapids section (IRS) of the river, leaving the development of the seaway to the two federal governments to resolve and pay for.

But this new "power priority" plan would nullify almost all of the agreed provisions of the failed 1941 Executive Agreement that coupled power and navigation. And while legislative approval would not be required, the US Federal Power Commission and its Canadian counterpart would still have to approve the plan, along with the all-important International Joint Commission. Truman still saw the two projects, the seaway and hydropower, as necessarily conjoined, perhaps out of concern that the seaway project, which he considered more critical, would be permanently sidelined if the "power priority" plan went forward. Shortly after winning the 1948 election, Truman publicly insisted that the coupled projects must advance together, and so all progress was frozen until Congress gave approval.

As well, one of the Federal Power Commission's examiners expressed the opinion that only the federal government, not New York, should develop power from the St. Lawrence. Although politically the FPC was nominally neutral, that opinion was seen as a political manipulation by the Truman Administration, and so at the start of 1950, the Power Authority submitted an appeal to the FPC. At the same time, the United States and

Canada actually did sign a treaty allowing for the mutual use of Niagara Falls, so long as it did not affect the beauty of the Falls. Thereupon Canada began work there on the new Sir Adam Beck 2 power plant.

A report to Congress from the US National Security Resources Board, echoed by several major US corporations, determined that newly discovered iron ore in the Canadian province of Newfoundland and Labrador would be indispensable to US military efforts in the event the Cold War intensified and, equally, that the seaway would be essential for transporting the ore, by way of Québec. The US steel industry needed 130 million tons of ore annually, and the leading US mines around Lake Superior and the mountains of Minnesota were being depleted at alarming rates. Truman concurred with the report, wrote Macfarlane, and promised that, in his presentation to Congress, the project would be "'dressed in uniform'—that is, to frame the project primarily as a military measure." The Truman administration, hoping for success, aimed at "impeding a separate [Canadian] power development and Canadian waterway, at times, misinforming Ottawa about the bill's actual progress.... Canadian patience with American delay was wearing thin." Those delays were "the biggest and longest dragging of feet I have known in my entire career," fumed the Canadian Secretary of State for External Affairs, Lester Pearson, soon to become Prime Minister. And so the plan for an all-Canadian seaway began to take shape and gain popularity. By now the burgeoning auto industry was providing all the business that the railroads could handle, almost eliminating resistance to the plan.

At about this time, the Power Authority's Board of Trustees elected a new chairman. John E. Burton had served as vice-president of Cornell University and state budget director. Thanks to the negotiations of his predecessor, Major General Francis B. Wilby, in June 1950 NYPA signed an agreement with the federal government, purchasing a 78-mile transmission line from Taylorville to the aluminum industry in Massena, built during World War II as part of the war effort. At long last, NYPA would begin earning its first revenue by leasing the facility to private utilities. Although the transaction was not completed for another year, it was a very important step, partly because the Authority's first operating asset was a transmission line, not a powerhouse, asserting its hard-won right to go beyond electrical generation. NYPA's 1951 Annual Report noted, "For the first time in the 20 years of its existence, the Power Authority became an operating concern with tangible property in its custody." And on March 20, 1951, in a move that seemed more aspirational than practical, the Power Authority Act was amended to include hydroelectric development of the Niagara River as well as the St. Lawrence River.

By then the Korean War had erupted—a proxy war between the United States and its allies, including Canada, against Communist Chinese troops armed with Soviet weapons. The war could only delay further any prospects for the St. Lawrence project. Worse, in December 1950 the Federal Power Commission rejected the Power Authority's license application, insisting that both the hydro and seaway projects should move forward, and that NYPA should share the new power with adjoining states. It was strongly suspected that the "independent" FPC was being influenced by the Truman administration. Once again, it appeared the St. Lawrence project was dead, and north of the border the all-Canadian project seemed the only viable option.

By this time both countries had identified the St. Lawrence project as the greatest obstacle in their relations. The Toronto *Globe and Mail* declared "There is no doubt that Canada can handle the project alone.... Ottawa should lose no time in making such a decision known," and other papers across the country echoed that sentiment. By December 1951 the governments of Canada, Ontario, and Québec were fully in accord, and they submitted their plans to the British Parliament for approval. There, the House of Commons would sit upon the proposal for more than two years before reaching full agreement with the United States, which was still rescuing Britain from harsh, postwar austerity with the Marshall Plan.

Nonetheless, Canada proceeded. "Going ahead with the seaway," as Canadian Minister Lionel Chevrier later wrote, "was notice to the US that we are now a first-class power. Some of my colleagues felt it was high time the US was made aware of this." This was a bold undertaking, given that the population of Canada at the time was less than 14 million, hardly more than the greater New York City region.

Canada's Liberal MPs were playing a tricky game. Their go-it-alone plan would still need US concurrence before the International Joint Commission for some aspects of the project, and so they tried to appear cooperative with Truman's efforts to bring Congress to approve the ten-year-old Executive Agreement and to revive NYPA's application to the Federal Power Commission. At the same time, a number of Canadian Members of Parliament said frankly that they hoped their project would "prevent the Americans getting their hands on the seaway at all," and were delighted "to see the US being shoved out of the limelight." Public opinion supported the Liberals' policies. Says Macfarlane, "The hope was that the threat of Canadian action and the sanction of the IJC [against the "power priority" plan] would lead enough recalcitrant members of Congress to approve the 1941 Agreement." Truman bluntly threatened Congress with this eventuality: "The question before the Congress, therefore, no longer is whether the St. Lawrence Seaway should be built. The question before the Congress is now whether the United States shall participate in its construction, and thus maintain joint operation and control over this development which is so important to our security and our economic progress."

In April 1952, the logjam of government entities with disparate goals approached its resolution. Truman finally agreed to joint power development, while Canada would proceed with its seaway alone. The President also allowed NYPA to be the designated American instrumentality to develop St. Lawrence hydropower and, in a meeting with Canada's new Prime Minister, Lester Pearson, agreed to submit the US power application to the IJC together with Canada's, after lengthy negotiations to smooth out differences in the applications. Since the Federal Power Commission had not approved the Power Authority's application, the application did not specify NYPA as the US developer. All the documents were submitted to the IJC on June 30, 1952. On Oct. 29, 1952, a week before General Dwight D. Eisenhower won the presidential election, the IJC granted permits for the power project to the Power Authority and Ontario Hydro. But how this enormous project would actually proceed still depended on whether or not there would be a seaway as well.

The Wiley Bill

In May 1953 the St. Lawrence Seaway Bill was submitted to the Senate by Wisconsin Senator Alexander Wiley, thus known as the Wiley Bill; a similar bill was introduced in the House. It was significantly less ambitious than the failed 1941 Executive Agreement and represented 20 percent of the cost—a mere $100 million. Despite these changes, wrote historian Austin W. Clark, "it was, in effect, sliding the United States neatly back into the terms agreed to, but never acted upon, with Canada thirteen years before.... [It also] drastically cut down on the scope of the project." Significantly though, the Wiley Bill withdrew the hydropower project from the federal government's plan. "The onus for this project," Clark continued, "would instead rest on the state of New York, which was more than ready to accept it. The Power Authority of the State of New York had already drawn up plans to work with the Canadian province of Ontario in a joint development."

Wiley described the situation succinctly to General Eisenhower during the latter's presidential candidacy in 1952. "Canada has offered to build the whole seaway, provided the Federal Power Commission grants a license to the state of New York to complete the power dams." Naturally, the bill was supported in both houses by the Midwest, but opposition remained fierce, predictably enough, from essentially the same political forces that Al Smith had encountered thirty years earlier. A bloc of legislators from coastal states remained unyielding in their opposition; these included Maryland, Virginia, Louisiana, Delaware, Georgia, New Jersey, New York (the downstate region) and Massachusetts (where freshman Senator John F. Kennedy bucked the trend and supported the bill).They were all in tune with the well-paid lobbyists of the Association of American Railroads, who objected most vociferously, describing the bill as "the nose of a camel under the tent," insisting that "this hundred million dollars would be but the first installment on a project which in its final form would cost many times that amount."

Another major objection that was hard to shrug off was that 90 percent of the US merchant fleet could not pass through such shallow canals. One senator told President Eisenhower that "he could favor it if ever he could be shown how a ship with a 34-foot draft could get through a 27-foot channel!" The opposition took this as proof that the bill was the "camel's nose"; doubtless the St. Lawrence promoters planned to increase the draft of the canals later on at an enormous added expense. So, by 1953, it seemed the bill was a lost cause and that Canada would soon begin dredging its own seaway.

That summer, the Federal Power Commission at last granted a fifty-year license for the St. Lawrence project as sought by the Power Authority, perhaps finally succumbing to political pressure in light of Cold War military concerns. With that, President Eisenhower issued an executive order allowing New York's Power Authority to construct the works with a foreign power. This culminated the twenty-two years of political efforts, since FDR had created the Authority. It would also allow the Canadians to begin their seaway immediately without American participation, as they were genuinely prepared to do by this time.

At the start of 1954, the last missing piece of the puzzle was the Wiley Bill, indispensable for the project to be binational. But obstruction from the same well-funded opponents was certain; the private power industry, railroads, the coastal and Mississippi shipping businesses, and all the unions that worked for them remained steadfastly against the project. After decades of successful obstruction, what could possibly make the difference in 1954?

The answer, simply enough, was President Eisenhower.

Eisenhower Seals the Deal

Dwight D. Eisenhower was sixty-three years old when he took office in January 1953, and arguably the most trusted man in the industrialized world, thanks to his meticulous military planning and execution as Supreme Commander of the Allied Forces a decade earlier. As chairman of the Joint Chiefs of Staff in 1946, he had supported the seaway in vain, so his support was taken for granted. But in his capacity as President he felt obliged to reevaluate the project and hear from all sides, including its most ardent critics in Congress, the railroad industry, and the Chicago Chamber of Commerce, all of whom reinforced Eisenhower's major concern: its cost. According to historian Clark, "In all circles, Eisenhower repeatedly said that he 'did not want to see us involved in any billion-dollar project on our own.'" But his Cabinet was primarily concerned with the possibility of an all-Canadian seaway, while the National Security Council focused upon the need for the iron ore from Newfoundland and Labrador, since World War II had mostly depleted US reserves of high-grade iron ore. And, as Clark reports, "It was also noted in a memorandum from the Canadian government ... that over ninety percent of the nickel used by the United States came from Ontario, specifically the area most in need of St. Lawrence power."

Ultimately, it was the security issue that won Eisenhower's approval for the Wiley Bill. As he later explained in his autobiography: "To me, this fact was central and overriding. Moreover, in time the St. Lawrence Seaway would become an economic necessity. The United States should move now, I was convinced, before a national crisis forced it into a crash program at a much greater cost." And by 1954, the opinion of the Eighty-Third Congress had also evolved, especially among Republicans seeking reelection that November, for whom the Wiley Bill became a priority, even as it sat stewing in committee where it was in danger of remaining.

By then the biggest difference was the forceful leadership of President Eisenhower. The St. Lawrence project had been supported—for the seaway, the power generation or both—by most of the last ten presidents, yet none had succeeded in swaying Congress. But now, after a year of fresh congressional hearings organized by the President to evaluate the project, the legislators were well-primed. So when the indispensable hero of the Second World War declared that the St. Lawrence Seaway was crucial for

US security, the impact was profound. Clark writes that "The centrality of iron ore to the issue of security cannot be overstated.... The need for iron ore to feed the great industrial machine of the Midwest was at the front and center of any writing on the subject."

Senator H. Alexander Smith of New Jersey, a longtime opponent of the project, changed his mind and spoke for millions of Americans when he declared that "there is no living man today whose views on our national security I would respect as much.... I must support the President of the United States in this field in which he is so pre-eminently expert." With some serious arm-twisting by Eisenhower, several others followed that senator into the pro-seaway camp. On January 20, 1954, the Wiley Bill passed the Senate with an equal number of Democratic and Republican votes.

In the House of Representatives, where the Seaway Bill was known as the Dondero Bill (after its sponsor, Michigan Representative George A. Dondero), the challenge would prove greater because of the gauntlet of hostile committees that used stalling tactics, trying to kill the bill's momentum until the midterm election. Although Eisenhower remained thoroughly engaged with House leaders, the outcome remained uncertain even on the day of the full House vote; but the Wiley-Dondero Bill passed handily, 241–158. One week later, on May 13, 1954, President Eisenhower signed the St. Lawrence Seaway Act into law, establishing the St. Lawrence Seaway Development Corporation to build the US portion of the seaway in conjunction with the Canadian effort. The hydroelectric plant was built jointly by New York's Power Authority and Ontario Hydro, without any clear division of responsibilities or costs. By that time Governor Dewey had appointed a new chairman of the Power Authority, a commanding figure who would be responsible for constructing NYPA's St. Lawrence hydropower plant, and who was arguably better qualified for the undertaking than anyone alive. His name was Robert Moses.

CHAPTER SEVEN

ROBERT MOSES PRODUCES PUBLIC POWER

The Greatest Achievements of Robert Moses

In 1981, *The New York Times* offered this obituary, attempting to sum up a career that defied abbreviation:

> Robert Moses was, in every sense of the word, New York's master builder. Neither an architect, a planner, a lawyer nor even, in the strictest sense, a politician, he changed the face of the state more than anyone who was. Before him, there was no Triborough Bridge, Jones Beach State Park, Verrazano-Narrows Bridge, West Side Highway or Long Island parkway system or Niagara and St. Lawrence power projects. He built all of these and more. Before Mr. Moses, New York State had a modest amount of parkland; when he left his position as chief of the state park system, the state had 2,567,256 acres. He built 658 playgrounds in New York City, 416 miles of parkways and 13 bridges. But he was more than just a builder. Although he disdained theories, he was a major theoretical influence on the shape of the American city, because the works he created in New York proved a model for the nation at large. His

> vision of a city of highways and towers—which in his later years came to be discredited by younger planners—influenced the planning of cities around the nation....

Moses's achievements are the stuff of legends, thanks mostly to the meticulous, encyclopedic research that Robert Caro conducted for his 1,200-page, 1974 biography, *The Power Broker: Robert Moses and the Fall of New York*. But legends have monsters as well as heroes, and Caro's chronicle depicts its subject as both. Caro was largely responsible for discrediting Moses by providing the harshest, most well-informed critique of his imperious methods and apparent disregard for much of the population he served, from whom he brooked no opposition. But Caro's assessment was not the most hostile; that came from the perspective of the Tuscarora Indians of Niagara.

In the decades since Caro's authoritative book appeared, Moses's reputation has sprung back a ways from the depths to which it was cast well before his death in 1981. And while Caro's overall evaluation is unquestionably severe and thoroughly documented, he insisted that his treatment was even-handed. In his introduction he observed that "Moses himself, who feels his works will make him immortal, believes he will be justified by history, that his works will endure and be blessed by generations not yet born. Perhaps he is right. It is impossible to say that New York would have been a better city if Robert Moses had never lived. It is possible to say only that it would have been a different city." Moses in return described Caro's work as "full of mistakes, unsupported charges, nasty, baseless personalities and random haymakers thrown at just about everybody in public life."

It may be a measure of the builder's expansive career that Caro devoted less than twenty pages to Moses's two most massive construction projects, on the St. Lawrence and Niagara rivers. But it seems likely that the construction of these two enduring, dynamic monuments did not fit the overall characterization of Moses that Caro chose to present. Either way it is unfortunate, because the two hydropower plants remain arguably Moses's greatest achievements, and that fact far outweighs anyone's opinion of his character.

Though born in Connecticut in 1888, Moses was a New Yorker through and through. Raised near Fifth Avenue in Manhattan, he studied at Yale and Oxford before earning his Ph.D. in political science from Columbia University. An interest in political reform led to his bold plans for reorganizing the New York State government, which by 1920 had brought the young man to the attention of Governor Al Smith, who gradually began seeking his advice, and with his reelection as governor in 1922, Smith brought Moses to Albany.

In 1921, the first state public authority, the Port of New York Authority, was created in the mold of its British precedent, the Port of London Authority. The Port Authority was "multipurpose," working on a variety of projects, each of which was financed by a separate bond issue and did not rely upon taxpayer dollars. That sort of pseudo-political model would well serve Moses's plans. In his recommendations for reorganizing the state government, Moses addressed the state's far-flung, isolated public parks and proposed consolidating them systematically into eleven regions, starting with Long Island, where he established a Long Island State Park Commission. He would serve as its president, of course, while controlling the development of Jones Beach State Park as well. All of that sounded politically harmless. But Moses also drafted the legal framework for a State Council of Parks, composed of all eleven park commissioners, which he would personally control with an iron fist and the unwavering support of Governor Smith. This allowed him to override the plans and projects of local park leaders across the state. In the mid-1920s, for example, he rode roughshod over the long-standing plans of the Niagara Falls Park Commissioners, for whom that sprawling parkland was part of civic life. Moses's imposition there was not forgotten thirty years later when he returned to build the Power Authority's Niagara plant.

After Moses redrafted the legislative mechanics of the "public authority," this unlikely new instrument (or more precisely, instrumentality) became his elevator to unprecedented political power. Then, in 1935, the very nature of a public authority was first expanded, not by Robert Moses, as Caro explains, but by Julius Henry Cohen, the general counsel of the

Port Authority, who, deep in the Depression, persuaded bankers to accept a consolidation of all of that Authority's separate projects' bonds into a "General and Refunding Bond." This allowed, for example, the profitable Holland Tunnel to bail out unprofitable bridges like the Henry Hudson. Cohen's innovation, born in desperate times, opened the door for Moses to create ongoing funds that could finance other projects; in fact, pretty much any project that he took a shining to. He only ran for office once (unsuccessfully, for governor in 1934) and yet through the creative use of the state's authorities under his control, for forty years he held sway over New York in a way hardly any elected official could match. In an especially telling passage, Caro notes a transformation in Moses at about this time. "Power was now, for the first time in his life, becoming an end in itself ... he was beginning to crave it now not only for the sake of dreams but for its own sake.... He was no longer satisfied with much of the power ... he now wanted *all* the power in the field of parks."

By the time Governor Dewey appointed Moses to the Power Authority in 1954, with the expectation that the trustees would elect him as chairman, Moses had more experience in large-scale construction projects than anyone alive. That fact alone gave him the most indispensable talent for construction of the St. Lawrence project: his proven ability to sell the bonds that would finance the huge facility. It is difficult to imagine any builder other than Moses who could have successfully attracted $335 million in long-term financing solely through the sale of bonds, without the use of any taxpayer dollars or state credit. True, the bonds that the Power Authority sold did benefit from the fact that they were not taxable, an advantage often bemoaned and decried by Con Edison and the other investor-owned utilities. But what controversy there was about financing centered around Moses's deal with Alcoa, the first of NYPA's St. Lawrence customers, whereby the aluminum company was awarded NYPA's first contract. Within a few years, according to historian Daniel Macfarlane, Alcoa had "contracted for about one-quarter of the power from the Moses powerhouse, and Reynolds Metals and a General Motors plant relocated to sites near Massena and signed power supply contracts, giving these three industries over half of the US power from the St. Lawrence development."

As general manager for the project, Moses chose Colonel William S. Chapin, who had helped build the Burma Road during World War Two before becoming Moses's chief aide as a consulting engineer for the Triborough Bridge and Tunnel Authority and the Office of New York City Construction Coordinator. Moses knew there was no question of hiring a federal organization like the Army Corps of Engineers, so he chose the private engineering firm of Uhl, Hall, and Rich for design (and compliance with the plethora of state, federal, and international laws pertaining to rivers, power plants, etc.), and the firm of Merritt, Chapman, and Scott for construction.

The division of the costs and responsibilities for the construction between the United States and Canada was settled with a handshake. "A joint meeting with the Hydro Electric Power Commission of Ontario followed," wrote Moses in his autobiographical *Public Works: A Dangerous Trade*. The chairman of Ontario Hydro, Robert H. Saunders, was responsible for the Canadian side of the project. "On my insistence and in the face of considerable skepticism on the part of the staff, a schedule of very rapid construction was agreed upon ... with first generation and the sale beginning in four years. Ontario Hydro and the New York Power Authority were to proceed with construction without written agreement but with an understanding based upon mutual respect. The work was to be divided on an equal basis, and a monthly transfer of funds was to be made covering the work performed. This unprecedented arrangement resulted in completion in record time."

Construction was expected to last as much as six or seven years, but on August 10, 1954, the day it began, Moses declared that the power project would be finished two years ahead of schedule. That was a bold pronouncement, given the geological uncertainties and engineering conundrums of such a large-scale project—and this was gigantic, as Caro eloquently described:

> There is the St. Lawrence—and, stretched across it, one of the most colossal single works of man, a structure of steel and concrete as tall as a ten-story apartment house, an apartment house as long as eleven football fields, a structure vaster by far than any of the pyramids, or, in terms of bulk, of any six pyramids together, a structure so vast that the thirty-two bright-red turbine generators lined up on its flanks, each of them weighing fourteen tons, are only glistening specks against its dull-gray massiveness. And this structure, a power dam, is only the centerpiece of Robert Moses's design to tame the wild waters of the St. Lawrence, a design that includes three huge control dams built to force the river through the power dam's turbines. After the dams were built—and the steel forests of transmission towers which distribute the electricity created by water passing through turbines—Robert Moses adorned their bulk with a garland of parks, of campgrounds, picnic areas, overlooks, of beaches built beside lakes that he built, and of miles and miles of more parkways.... [Soon after] at Niagara, Robert Moses built a series of dams, parks and parkways that make the St. Lawrence development look small.

The magnitude of the entire project on the St. Lawrence is difficult to grasp. In all, 22,000 workers from the United States and Canada worked on the combined power dam and seaway project, more than half of whom were brought in from other regions. "This is better than the circus," said one construction worker. Digging for the seaway produced vast quantities of excavated material—more than 27 million cubic yards—that had to be relocated on the US shore. Although overall the safety record for the project was good, relative to its scale, nonetheless 42 deaths on the site were reported and, Macfarlane reports, "scores of people were seriously injured, and accidents such as cables mangling or severing legs led some workers to call it 'Cripple Creek.'" The men kept working through adverse weather, even in January 1957 when the temperature sank to -57°F.

Shortages of specific supplies caused some delays, as did several strikes. Even six months before completion, it was uncertain whether the demanding schedule would be met.

To finance the US construction project, on December 1, 1954, the Power Authority issued bonds for the first time, totaling $335 million at 3.18 percent interest. And a little over a year later Governor W. Averell Harriman approved the first contracts for the St. Lawrence Power Project to sell power to the City of Plattsburgh municipal electric system, Plattsburgh Air Force Base, and the state of Vermont. Sales to other neighboring states would follow, as required by the federal license.

Challenges of the St. Lawrence Project

In spite of the friendly trust he had established with Saunders, Moses's autocratic leadership throughout construction was duly noted by everyone engaged, including Canada's team, with whom significant chafing arose. Over the previous two decades, many Canadians had become convinced that an all-Canadian seaway was an essential national priority, and so at various points along the river, Ottawa insisted that the dredging conform to plans that could facilitate a separate Canadian seaway later on. And ambiguities in the original approved plans left many details in dispute. For example, "the US government, with the major exception of [NYPA]," writes Macfarlane, "was firmly of the opinion that the power entities were responsible for excavations downstream of the dam." Other parts of the preliminary plan proved impossible to carry out and had to be adjusted, often at a greater cost than anticipated. Disputes of this kind were ultimately reduced to monetary divisions between the two nations, and more often than not, Robert Moses had his way. One Canadian expressed the opinion, shared by many, that Moses was "a quasi-dictator who treated the park—and the power authority—as his own personal fiefdom, but he got things done, and usually got them done well."

The most difficult issue was land acquisition, which sometimes meant the forced relocation of residents on both sides of the river where the new

Lake St. Lawrence would be created by the dams, in some cases wiping out communities and orchards. The New York Department of Public Works was responsible for negotiating settlements. This was especially difficult with Mohawk communities, including more than 6,000 residents of the Akwesasne (or St. Regis) reservation, which spread into both countries and claimed Barnhart Island, the site of the Moses-Saunders Power Dam, as it was formally named. Far fewer residents were displaced on the US riverside than on the Canadian, where some 8,000 people were relocated from what came to be known as the "Lost Villages" in Ontario and Québec. In the United States, as late as 1957 some two hundred Native Americans occupied part of a broad creek near Fort Hunter to protest their eviction, but for naught. Moses himself declared that their sacrifice was necessary "for the common good," but warned that the Power Authority, which was responsible for the displacements, would not be "cajoled, threatened, intimidated or pressured into modifying sound engineering plans to suit selfish private interests"—if wishing to remain in their homes could be so described. Moses's attitude contributed to deteriorating relations with the Native American tribes that would resurface later.

Part of the challenge of the combined St. Lawrence project lay in the competing interests for the regulation of the river's flow: those concerned with flood control wanted moderate water levels, while the shipping industry and power station operators sought to maintain higher levels. To regulate the flow above and below the dams, the International Joint Commission created the International St. Lawrence River Board of Control in 1956.

Despite the severe Northern New York winters, Moses's accelerated construction schedule did indeed hold, with the start of electricity production on July 17, 1958, a year before the seaway was completed. On June 27, 1959, Vice President Richard M. Nixon joined Queen Elizabeth II in formally dedicating the power project as a symbol of international cooperation. The Queen unveiled a monument at the power dam, attesting to "the common purpose of two nations whose frontiers are the frontiers of friendship, whose ways are the ways of freedom, and whose works are the works of peace." Full power—1,600 megawatts equally divided between New York and Ontario—was delivered on July 20, 1959, fully two years ahead of the original schedule, exactly as Moses had avowed.

The American Society of Civil Engineers named the St. Lawrence project the most outstanding civil engineering achievement of 1959, and in 2000, the American Public Works Association rated it among the ten most important public works projects of the century. More significantly, after decades of political struggle for public waterpower in New York, the thread of determination from governors Hughes, Smith, Roosevelt, Lehman, Poletti, Dewey, Harriman, and Rockefeller at last paid off. In 1958, that determination began rewarding the people of New York State.

Tragedy Strikes

Midway through the St. Lawrence construction, the future of New York's power generation was transformed by a catastrophic event three hundred miles to the west. As later reported by the *Niagara Gazette*, "When the Schoellkopf Power Station collapsed into the Niagara Gorge on June 7, 1956, it wiped out in seconds nearly 25 percent of the city's tax base," delivering "a devastating blow to the area economy and an ironic end to the debate over public versus private power development," given that tens of thousands of jobs in the region were eliminated in an instant. More than 60 percent of the station was reduced to "twisted girders, rock and rubble in the lower Niagara River. Three 70,000-horsepower and three 32,500-horsepower generators went into the river." A 25-year-old plant operator recalled seeing "a crack open up ... and I made tracks, headed down the floor to the north. I thought maybe I could get the elevator up, but it was too late.... There were rocks falling behind me, and I was running. The water was chasing me down the floor." One worker was killed.

This plant was now owned by the Niagara Mohawk Power Corp., the largest municipal taxpayer in the booming industrial city of Niagara Falls, and it was the only US power plant. Producing 450,000 kW of power, it had often been dubbed the world's greatest hydroelectric plant. Company executives' estimates of the damage ranged from $20 million to $100 million. But the broader economic impact was much greater, constituting, in fact, a regional emergency. A patchwork system of short-term solutions was improvised statewide to provide limited power, but the situation was untenable without a major new source of electricity.

Two days after the rock slide, Rep. William E. Miller (R-Lockport, NY) asked that Congress act with urgency to facilitate the reconstruction of Schoellkopf, insisting that the rightful group for the job "certainly should be the private utilities since no other agency has developed any power on the American side of the river" (which was technically true, but only because the St. Lawrence project was still under construction). In 1953 the Republican-controlled House had passed a private-development bill, but Governor Dewey, though a Republican, followed the course of every New York Governor since Al Smith and maintained his commitment to public development of Niagara and the state's other waterpower resources; with that, the US Senate rejected the private power bill. Then New York's former Democratic governor and now senator, Herbert Lehman, promptly sponsored a bill providing for public development of Niagara hydropower by NYPA, which Republicans just as promptly opposed, despite Moses's unequivocal declaration: "The basic underlying issue is the unquestionable public ownership and inalienability of the greatest natural resource of the state of New York."

Before long, however, officials of Niagara Mohawk had to acknowledge that it was impossible to rebuild a powerhouse at the Schoellkopf site, and Congress, responding to pressure from the White House, passed the Niagara Redevelopment Act, which Eisenhower signed into law on Aug. 21, 1957. It directed the Federal Power Commission (FPC) to issue a license to the New York Power Authority to redevelop Niagara Falls' hydroelectric power. The following January, the FPC issued a 50-year license to the Power Authority to build and operate the Niagara Power Project.

Tackling Niagara

Six weeks later Moses launched the construction, which would ultimately cost $737 million, more than double the US cost of the St. Lawrence project. He proclaimed that, with crews working around the clock, power would flow from Niagara in less than three years, which gives some idea of the urgent need for power. For design and compliance, he chose the same private engineering team—Uhl, Hall, and Rich—that was still completing the Moses-Saunders Power Dam; and the prime contractor for construction was once again Merritt, Chapman, and Scott. The Power Authority employed an army of 11,700 workers, temporarily swelling the population of Niagara Falls to a record 102,000 in the 1960 national Census.

Robert Moses was, of course, the singular general of this army. One of its foot soldiers was a local eighteen-year-old named Ken Glennon, who signed on early in the project. He served as a "field checker," a timekeeper who tracked the workers and their hours across most of the site's many square miles, which gave him personal knowledge of the enormous, determined crew that got the job done. Glennon chronicled the work of the men and women who put out that effort in a book, *Hard Hats of Niagara*, published in 2011, on the fiftieth anniversary of first power. Glennon's collective account is truly "history from the bottom up," recounting hundreds of anecdotes from dozens of engineers, contractors, residents, nurses, and their families. They depict in detail the enormity, complexity, and difficulty of constructing the project, which resulted in twenty fatalities and myriad life-changing injuries. Each crew member had to be smart and strong: much of the equipment, like the enormous Caterpillar D9 bulldozer, was not yet hydraulic, so the blade was raised and lowered by hand, using pulleys; a crescent wrench was often the only "switch" for adjusting the controls.

What is most apparent from their stories is that, when they reflected back upon it, for many this project represented the pinnacle of their professional lives, despite many real hardships for the workers and their families—for example, the absence of available housing within fifty miles, even in trailer parks. Nonetheless, dozens from this enormous group still held reunions twenty-five and fifty years after completing the project. Construction crews don't typically hold reunions, but the Niagara project was different. Even fifty-five years later, in 2015, scores of workers returned to the Power Project to socialize with their old friends and re-share many of the tales and photos that appear in Glennon's book. What shines through all the stories is an exceptional esprit de corps, a camaraderie clearly rooted in the fact that their collective work was both critical and urgent for the people of the state and not merely a matter of profit for investors. Those whose deaths they memorialized at these gatherings reminded them of the gravity of the situations they had endured. And their pride in being personally responsible for a small part of this unique, monumental project is undeniable, even among those former workers who had been fired (usually for safety violations or the loss of valuable equipment).

The sheer scale of the project often left them dumbstruck. For example, the two trenches for the conduits, each four miles long, that diverted the river to the forebay, "were over one hundred feet deep and sixty feet wide," wrote Glennon, "and blasted from solid rock for the entire length." After the trenches were blasted, the cement lining was poured their full five-mile length in forty-foot sections along the leveled walls and floor, even as huge rocks and small boulders periodically fell from high above, to warning shouts of "Headache!" Then cement finishers, like Joe Sardina, went to work "down in the hole," after which the entire conduit was covered. "Once the section was finished," Joe said, "it was awesome to stand at the bottom and look up at the arched ceiling over sixty feet above." The conduit "seemed to extend to infinity," he added, "as if it were an enormous cathedral chamber of technological achievement. Visualizing this titanic chamber flooded with cascading Niagara River water defied comprehension."

Hardly any of the hard hats had a clear understanding of what all the other crews were doing across the hundreds of acres, but each phase of the construction was meticulously planned and timed for one to follow another. The blasting came first, then the loose rock was crushed down to pebble size and added to the concrete mix, which was poured into the wooden frames where the skeletal rebar was in place, then smoothed by the finishers even as giant equipment was coming through to another workstation and temporary bridges were being erected overhead. This went on twenty-four hours a day, seven days a week.

Keeping Niagara Beautiful

As a world-renowned tourist site, Niagara Falls received very serious aesthetic consideration. An average of more than 200,000 cubic feet of water per second (or 1.5 million gallons a second) flows from Lake Erie into the Niagara River, where 50,000 to 75,000 cubic feet per second would be diverted 2.5 miles above the Falls into tunnels leading to the power plant. That is 600,000 gallons of water per second rushing through the twin conduits, 46 feet wide, running four miles to the project's forebay, where it would enter the thirteen turbines through 460-foot-long penstocks, 28.5 feet in diameter, and then discharge into the Niagara River. The turbines were rated at 200,000 horsepower each, with a total output of 1,800 megawatts (since then raised to 2,675 MW).

The diversion could reduce the drama and scale of the Falls, and the 1950 treaty between the United States and Canada specified that at least 100,000 cubic feet per second of water must spill over the Falls by day during tourist season (April through October) to maintain an "unbroken curtain of water"; that flow could be reduced by half at night and the rest of the year. That is why the design called for a reservoir holding 22 billion gallons of water to feed the turbines during daytime. Apart from that requirement, Moses had free rein, structurally and aesthetically, over the entire area, given his control over both the Niagara power project and Niagara State Park. In fact, the Power Authority committed $15 million of its power project budget to a new, eighteen-mile scenic parkway to and through the site, which has been a sore point with many residents ever since.

Legal Battles

To create the reservoir, residents who had lived on the same land for many hundreds of years would have to be displaced: this time it was 175 members of the Tuscarora tribe and their thirty-seven homes. Moses planned to flood 1,383 acres—one fifth of their land—in Lewiston. According to Ginger Strand's history, *Inventing Niagara: Beauty, Power, and Lies*, Moses's initial offer to the tribe was just $1,000 per acre, while at the same time he was offering adjacent Niagara University $50,000 per acre.

But Tuscarora resistance to Moses was far more effective than the defensive efforts of the Akwesasne had been at St. Lawrence. After raising his offer for Tuscarora land to $1,100 an acre and promising a $250,000 community center on their reservation "in the event that an early agreement is consummated," Moses was faced with an ongoing legal battle from the tribe after a Federal Court of Appeals rejected the builder's eviction procedures. The Indians' petition reached the US Supreme Court, challenging the validity of the Power Authority's license because it was predicated upon their unjust displacement.

In September 1958, Moses issued a statement entitled "On Tuscarora Obstruction," protesting the delays that were costing the project "$100,000 a day" because of the Native American legal tactics that he deemed "a silly game." "We have been shunted about and jackassed from court to court and judge to judge, and are faced with the prospect of more litigation and further delays.... I have been involved in quite a few public enterprises with the usual conventional, outrageous and also comparatively innocent pressures, but nowhere in my experience has there been so much of this as on the Niagara frontier." At the end of this outburst, Moses inadvertently exposes his political leanings. "How our democratic system can survive such stultifying domestic weakness, incompetence and ineptitude in the ruthless, world-wide competition with other systems of government more incisive and less tolerant of obstruction, is more than I can figure out." As the renowned and eloquent critic Edmund Wilson pointed out in *The New*

Yorker, "The implication would seem to be ... that our 'democratic system' can hardly survive without becoming less democratic."

Moses escalated his threats while sweetening his offer to $2.5 million, which the Tuscarora also rejected, insisting that the land was their mother and "you cannot sell your mother." The Federal Court of Appeals agreed with the Tuscaroras' contention that their land was not "necessary for the project but was desired solely for economy." Saving costs at the expense of the Native Americans' homes was never Congress' intention, said the Court, which left it to the Federal Power Commission (FPC) to certify that the project "would not interfere with the purposes of the reservation."

Moses insisted that the added costs would result in a 30 percent increase in the cost of the power produced, but he also raised his offer to $3 million, which the Tuscarora again rejected. Then, on February 9, 1959, to the surprise of many, the FPC ruled that the Tuscarora could not be compelled to give up their federally granted reservation against their wishes. Moses called the decision "gobbledygook." "I'm delighted," declared Chief Black Cloud of the Tuscarora. "I expected to get justice from the federal government, and we did."

But that was not the end. After further extensive hearings, the FPC determined that the added delay and cost of resituating the reservoir would cause greater community disruption and expense while diminishing the power that could be generated with the use of the reservoir, and thus would violate the Niagara Redevelopment Act. For that reason, the case went to the Supreme Court in *FPC v. Tuscarora Indian Nation*, 362 US 99 (1960).

On March 7, 1960, a majority of Supreme Court justices ruled in favor of the FPC on the grounds that the 1,383 acres in question did not constitute "reservation" under the legal definition: they had not been part of the original Tuscarora reservation but had been purchased by the tribe later for "fee simple," and had not yet been developed. Besides, the ruling noted, "It would be very strange if the national government, in the execution of its rightful authority, could exercise ... the power of eminent domain in the several States, and could not exercise the same power in a Territory occupied by an Indian nation or tribe, the members of which were wards of the United States, and directly subject to its political control." Justice Hugo Black wrote for the three dissenting justices and described the definition of "reservation" as a technical triviality and the ruling as a "broken promise," insisting that "great nations, like great men, should keep their word."

Ginger Strand writes that, "like the explorer LaSalle, Moses arrived at the Niagara frontier with dreams of technological mastery in the service of greater use, and was stopped short by a completely baffling Indian worldview that seemed to have no interest in using the land for profit and no interest in letting him do so. Like LaSalle, Moses bulldozed his way across this worldview, pretending to be conciliatory while doing exactly what he wanted. He too left his name plastered all over the landscape.... For Moses, as for nineteenth-century advocates of Indian removal, the Indians' failure to *do* anything with the land proved that they didn't deserve it."

Completing the Niagara Power Plant

Within three years, 12 million cubic yards of rock were excavated. The massive main structure was 1,840 feet long, 580 feet wide and 384 feet high. Eight temporary bridges were built to accommodate the rerouted traffic on both roads and railways. A church was moved and seventy-four houses were relocated at the rate of two a day to a new residential development three miles away.

The thirteen generators were built by Westinghouse Electric Company more than half a century after Nikola Tesla and George Westinghouse installed the first AC generator at Niagara Falls. Since each unit weighed more than a thousand tons, they had to be assembled after being lowered in separate parts. It was especially difficult in the dead of winter, of course, which contributed to the twenty deaths on the job. The temporary dirt haul-roads, along which even the heaviest cranes had to travel, were perilous; two 24-ton Euclid tractors, each worth $47,000, fell into the river. Despite the tragic deaths and countless setbacks and reversals, the project was completed two weeks ahead of its three-year deadline. At 2,400 MW (including Lewiston pumped storage), it was for a time the largest hydropower complex in the western world. More than a half century later, with the installed capacity raised to 2,675 MW, it remains New York's largest power project. In a good year, it saves consumers more than $737 million, the cost of its construction. That is partly thanks to the much-disputed Lewiston reservoir, which served as the single largest "battery" in the world and remains vital today for storing power.

The Lewiston Reservoir

Everyone with a cell phone knows the importance of a battery; storing electricity remains the greatest challenge in power management today. That's not surprising, given that electricity is dynamic in its essence, produced, as Michael Faraday first demonstrated, when a magnet transforms motion into an electric current. So it's not easy to trap such a lively spirit in a small tin cylinder for your flashlight or a square box for your car. Energy storage is the most urgent challenge for the power industry to surmount, because it will unleash the full benefits of such intermittent, renewable energy as solar and wind. Virtually the only way of harboring power on an industrial scale is with reservoirs that can store, not electrical energy, but gravitational potential energy.

In the late nineteenth century, the Swiss devised systems in the lower hills of the Alps and Juras whereby a river-fed upper reservoir would, upon the opening of a valve, release water through a turbine to a lower reservoir, thus providing electric power upon demand; then, when there was the least demand for electricity, a set of pumps pushed all that water back up to the higher reservoir. It is called a pumped-storage system, and in those early days it was highly inefficient, using far more electricity to drive the pumps than it could subsequently generate; but it produced power *when it was needed*. A plant near Niederwartha, Germany, made advances with the technology in the 1920s.

The first pumped storage facility in the United States was built in New Milford, Connecticut, in 1930. There, two 8,100-horsepower pumps lift water at night, when power is not in demand, from the Housatonic River to a reservoir 230 feet above. Then, whenever the state grid needs additional power, that water is released back into the river through turbines, driving a 44,000-horsepower generator. The first reversible turbines for pumped storage facilities were developed during the 1930s, capable of operating as both turbine generators and, in reverse, as electric pumps. These innovations and other improvements have raised the average efficiency of pumped-storage facilities to 80 percent today. Seventy-five years later, the Rocky River Pumped Storage Plant in New Milford is still owned and operated by the Connecticut Light and Power Company, and rated at 29,000 kW. According to the Electric Power Research Institute, pumped storage accounts for more than 99 percent of the world's bulk energy storage, or some 123,000 MW. What prevents more widespread use of pumped storage is the very specific topography required, with a source of water and space for a large reservoir above a steep drop to the power plant, and with a second reservoir or river below.

That was not a problem at Lewiston, adjoining Niagara Falls, where the site for the reservoir was almost ideal, which is exactly why Moses fought so hard to acquire it. The Lewiston Dam would contain the new 1,900-acre reservoir, from which water pours down through twelve turbine pumps during the day when demand is highest (this is called "peak shaving"), and then through the main Niagara powerhouse's thirteen turbines as well. At night, when demand for electricity and its price are both lower, the Lewiston turbines operate in reverse, as pumps, and convey water back up to the reservoir (called "valley filling"). The Lewiston plant can also provide power for sudden, unexpected demand (called "spinning reserves"). By providing additional flexibility in load balancing, Lewiston is an ideal complement for the Niagara Power Plant, and obtains all the power its pumps need from the main Niagara plant itself, rather than remote sources.

The Robert Moses Niagara Power Plant of the New York Power Authority was opened on January 28, 1961, and the federal government, which for decades had obstructed the project more often than supporting it, became its proudest godfather. An event at Niagara University on February 10, 1961, gathered some 4,500 guests for formal celebration of first power. Three former presidents—Eisenhower, Truman, and Hoover—recorded messages that were broadcast during the event. The voice of President Kennedy, still in his first month in office, heralded the project as "an outstanding engineering achievement" and an "example to the world of North American efficiency and determination." President Eisenhower was more eloquent. "The achievement," he said, "reflects enlightened international, national and state leadership and cooperation. The mighty power of the Niagara has been harnessed for the public good, and the beauty of historic Niagara Falls has been preserved for all time." Then governor Nelson A. Rockefeller, with Robert Moses at his side, symbolically flipped a switch and launched the first power.

CHAPTER EIGHT

JAMES FITZPATRICK AND THE NEXT GENERATION

The Changing of Hands

Robert Moses retired, inadvertently, at the end of 1962. Protesting governor Nelson Rockefeller's decision to replace him at the all-powerful State Council of Parks, Moses tendered his resignation from *all* his posts, a ploy that had served him well with several governors over the decades. But Rockefeller called his bluff and accepted all his resignations willingly and doubtless with relief. Moses was blindsided. According to a colleague of his, cited by Caro, "'I don't think Mr. Moses realized what he was doing in the rush that day. I don't think he really expected the Governor to let him resign [from all the park posts] and I don't think he ever expected him to let him resign from the Power Authority. When he did realize it, I think it broke his heart....' Once, he had held twelve separate posts," wrote Caro. "Now he was down to two," including the presidency of the World's Fair, coming up in 1964. "But the money came from his network of public authorities. Now, at a stroke, three of them were gone—including the biggest of them all, the State Power Authority that, with both Robert Moses Power Dams completed at last, was beginning to generate tens of millions of dollars in annual revenue." After almost nine years at the head of NYPA,

Moses was gone, and while no one better could have been chosen to build the NYPA facilities, with construction completed, he may not have been the best man to operate them. He was known as "the Master Builder" for good reason.

At the beginning of 1963, it fell to James A. FitzPatrick to succeed Moses and lead the organization into its second generation. Born and raised in Plattsburgh, New York, FitzPatrick earned a law degree before reaching the rank of Lieutenant Commander in the Navy during World War II and serving in the State Assembly from 1947 to 1956. His subsequent election as chairman just as NYPA came of age proved so felicitous that he remained in that post for more than fourteen years, longer than any other chairman in the authority's history. FitzPatrick "oversaw the acquisition and construction of almost $2 billion in new facilities," lauded his *New York Times* obituary in 1988. "The improvements made under his guidance more than doubled the generating power of the Authority ... which now provides about one-third of the electrical power used in the state."

Blackout

The boldfaced headline of the *New York Times* shouted "Power Failure Snarls Northeast; 800,000 Are Caught in Subways Here; Autos Tied Up; City Gropes In Dark; Snarl at Rush Hour Spreads Into 9 States." The front-page article described how, the day before, Tuesday, November 9, 1965, "the largest power failure in history blacked out nearly all of New York City, parts of nine Northeastern states and two provinces of southeastern Canada.... Some 80,000 square miles, in which perhaps 25 million people live and work, were affected." In fact, the lives of some 30 million were put on hold, though few fatalities were reported. The runway lights went dark at all the Northeastern airports just at sunset. Gas stations could not pump gas from their electric pumps. Those riding the subways were trapped and may have suffered the greatest trauma; they had to "make their way along tunnel catwalks, shepherded by employees with emergency lights," according to the *Times*. In one tunnel, a blind straphanger proved the most adept at guiding other riders along the tracks and up to the street. That night many UFO sightings were reported, as some city dwellers saw the stars against the black sky for the first time in their lives.

"Railroads halted," the *Times* went on. "Traffic was jammed. Airplanes found themselves circling, unable to land.... The lights and the power went out first at 5:17 P.M. somewhere along the Niagara frontier of New York state. Nobody could tell why for hours afterward. The tripping of automatic switches hurtled the blackout eastward across the state—to Buffalo, Rochester, Syracuse, Utica, Schenectady, Troy and Albany. Within four minutes the line of darkness had plunged across Massachusetts all the way to Boston. It was like a pattern of falling dominoes—darkness sped southward through Connecticut, northward into Vermont, New Hampshire, Maine and Canada." Most power was not restored until the next morning. Famously, nine months later the *Times* also reported that the birth rate spiked in the Northeast, since couples had no distractions during the blackout, a romantic notion later debunked by more level-headed demographers. It is true, however, that crime was at an all-time low that night and it was widely noted that during those fourteen hours New Yorkers showed an uncharacteristic degree of civility.

The cause of this massive blackout took a while to understand, especially because the companies responsible for generation and transmission in the Northeast tried to deflect the blame. It soon became clear that: a) any number of unforeseeable circumstances could trigger a catastrophic blackout; and b) the entire Northeast grid, which had evolved rapidly and haphazardly as new communities were added, was so vast and interconnected that it defied any comprehensive understanding. Nor was there any quick fix at hand.

According to the Canadian Broadcasting Corporation, at 5:16 p.m. "a single transmission relay failed leading from the giant [Niagara] Sir Adam Beck No. 2 Generating Station at Queenston, Ontario." It transpired that the cause was a human error one week earlier, when maintenance workers at that Canadian plant had lowered the setting too far on a circuit breaker, so that the slightest surge of power at the Robert Moses plant across the river would shut down the transmission line between the two. That triggered a cascade of failures as the US power, with nowhere else to go, raced through other live wires, overloading their circuits to the west and then to the east as load imbalances shut down other power plants along the way, producing total, incomprehensible chaos across the system.

It was all the more remarkable, then, that NYPA's two power plants and all its transmission lines remained intact and operational throughout the blackout. NYPA continued delivering power to the city of Niagara Falls and surrounding areas. In some cases NYPA's power made a crucial difference in stabilizing the system to relaunch the generators from a "black start" at immobilized private utilities. Municipal utilities with their own generation in New York, Connecticut, and Massachusetts were able to disconnect their systems from the grid and so were unaffected.

The Blackout of 1965 demonstrated that the electrical foundation for the most populous region in the industrialized world was stunningly fragile and vulnerable. A minor mishap in Ontario had brought almost all the interconnected plants as far as southern New Jersey screeching to a halt, and there was no reassurance it would not happen again, perhaps repeatedly and with lethal consequences. President Lyndon Johnson demanded a full explanation from the Federal Power Commission and put the resources of the FBI and the Department of Defense at its disposal. In its report, the Commission made 19 recommendations, mostly calling for better collaboration and communication with Canada.

One important outcome was the creation of the North American Electric Reliability Council (NERC) in 1968, with the mission to "ensure the reliability of the North American bulk power system," which today (still NERC, but the North American Electric Reliability *Corporation*) serves more than 334 million people. NERC divided the continental United States and Canada into ten "reliability regions." They include the region covered by the Northeast Power Coordinating Council, which stretches from southern New York to the northernmost reaches of Eastern Canada, assuring a balanced load within the region. (Of course, those who endured the more far-reaching Northeast Blackout of 2003 could be pardoned for their skepticism. The cause of that event, which crippled the activities of fifty-five million people, was stunningly banal: an unpruned tree in Ohio triggered a daisy-chain of transmission failures that shut down 265 power plants in the United States and Canada.

Another result of the 1965 Blackout was the creation of the New York Power Pool (NYPP), established two years later as a coordinating institution and centralized reliability organization for wholesale power. The Power Pool continued to serve as New York's operational center for the next thirty-two years, before evolving into today's New York Independent System Operator (NYISO). Created by the seven major investor-owned utilities in the state, the Pool provided them with both a statewide system monitoring for any lines or units that had tripped, as well as an automated wholesale market, whereby they could pool their operations for mutual benefit and reduce the collective capacity that they required. FitzPatrick sent representatives from the Power Authority as consulting observers at first, who served on several NYPP committees before NYPA became the Pool's eighth full-fledged member. According to one analyst, "NYPP carried out many of the reliability functions normally performed by a control area operator.... Above all, NYPP was a voluntary reliability coordination organization, with narrow but still important functions to balance generation with loads in a reliable fashion and to share whatever savings could arise from efficient system balancing." Since reducing their generating costs was a large part of the utilities' motivation, it is no surprise that the cost of creating and operating the NYPP was passed on to their customers.

Exploring Nuclear Energy

At the start of FitzPatrick's tenure, the Power Authority did not yet provide service to the people of New York statewide. In the absence of adequate transmission lines down to the populous southeastern counties, and in line with federal and state requirements, the hydropower went to upstate factories that provided thousands of jobs, to public municipal systems, and to rural cooperatives. In some areas, three upstate private utilities distributed NYPA's electricity to their residential and farm customers without making a profit themselves. And under federal mandate, a portion of the electricity was sent to neighboring states.

As demand increased across New York, FitzPatrick heard from more and more companies seeking Power Authority electricity with which to expand and create more jobs; the municipal and rural public power systems would also need additional electricity. Of course, there was no other New York river like the Niagara or the St. Lawrence, and while smaller hydropower plants were a possibility, they were no panacea. But from the perspective of the 1960s, another technological development seemed to be exactly that.

The first nuclear power plant began operating in the Soviet Union in 1954, generating just 5 MW of power. At the end of 1957, the first commercial nuclear power plant in the United States became operational near Pittsburgh, Pennsylvania; by 1970, although only one percent of US generation came from nuclear power, more than 85,000 MW of nuclear capacity was under construction or in the planning stages. And over the next decade, as petroleum costs rose an average of 26 percent, natural gas 23 percent, and coal 16 percent annually, nuclear power seemed much like a panacea to the world's energy needs. Until it didn't.

Commercial nuclear power is produced when neutrons bombard atoms of uranium or other radioactive material to unleash more neutrons, creating a controlled chain reaction that produces heat. Nuclear reactors boil water, which becomes the steam that drives the turbines to generate electricity. As the nuclear fuel rods exhaust their radioactivity, they must be replaced with fresh uranium, and the old rods must be disposed of with immense care in sealed pools or caskets, where they must remain for many hundreds of years to prevent radioactive contamination of natural surroundings and human inhabitants. All this to heat water.

Nuclear power was a contentious issue from the first, but its promise seemed irresistible. By 1967, FitzPatrick was eager for the Power Authority to add a nuclear plant within a practicable distance from the power-hungry New York City area. Legislation was drafted to permit the Power

Authority to build one nuclear plant and to proceed with a second pumped-storage hydro project similar to the one adjoining Niagara.

FitzPatrick set out on what would evolve into an ambitious, $3.5-billion series of construction projects that expanded NYPA's generating and transmitting capacity along with its fundamental responsibilities. As the only state-owned generating entity in New York, NYPA had an inherent obligation to anticipate and prepare for future demand by undertaking efforts that others could not or would not make. The private utilities were only responsible to their shareholders and customers, aiming to grow their corporations without competition from the Power Authority.

So even before the plan was formally unveiled by New York's two most powerful Republicans, Governor Rockefeller and Senate Majority Leader Earl W. Brydges, the Republican-aligned private utilities began objecting. Brydges managed to pass the bill in the state Senate but in the Assembly, where the private utilities had more clout, the legislation died in committee. The utilities might have compromised on the public pumped-storage proposal, but they were not prepared to allow the Power Authority to enter the nuclear power arena without a fight. At the same time, a number of senators were suspicious and dropped nuclear authorization. New York's Senator Robert F. Kennedy called the plan "a sellout" to the private utilities that protected their "monopoly" over nuclear power.

To break the stalemate and address the growing threat of future power shortages, Rockefeller selected eighteen members for a blue-ribbon panel to evaluate the state's long-term needs. It was chaired by Dr. R. G. Folsom, the president of Rensselaer Polytechnic Institute, and included NYPA's FitzPatrick, along with representatives of the private utilities, the municipal systems, rural cooperatives, and business and labor groups. In December 1967, the committee recommended authorizing the Power Authority to build pumped-storage projects as well as nuclear power plants, to serve both industries and the municipally-owned systems.

Just such a bill then passed the Legislature with bipartisan support. Rockefeller signed it into law in May 1968, describing it as "a unique partnership between government and private industry in meeting the future power needs of the state." By August, the Power Authority sought a license from the Federal Power Commission to build a pumped-storage generation plant near the Schoharie County towns of Blenheim and Gilboa, about forty miles southwest of Albany. NYPA also applied at the end of 1968 to the US Atomic Energy Commission for permission to build a nuclear power plant on the south shore of Lake Ontario, near Oswego, to serve "high-load factor industries" and to provide power to municipalities and rural cooperatives across the state. Less than a year later, construction began on both projects.

Ground-breaking ceremonies took place at Blenheim-Gilboa in July 1969, whereupon construction began with a workforce that peaked at 1,700. A total of some 7.5 million cubic feet of Earth and rock were removed and almost a quarter-million cubic yards of concrete were poured. When completed, the lower reservoir was filled with 5 billion gallons of water by impounding Schoharie Creek. A thousand feet above, on Brown Mountain, another reservoir the same size was dug out. When power is needed, the water from the upper reservoir travels down a tunnel 28 feet in diameter, through four Hitachi-built pump/turbine units to the lower reservoir, producing 260 MW each (since upgraded to 286 MW). At night, the same units pump the water back to the upper reservoir, using electricity from the grid to drive the pumps. Most of the project is underground, beside the lower reservoir; only the administration building is visible, along with the electric switchyard and a muscular, 510-ton gantry for raising and lowering the turbines when required. (In 2010, a $135 million life-extension and modernization upgrade was completed on all four turbine/pumps, ensuring the plant decades more of reliable service.)

On July 5, 1973, the Blenheim-Gilboa Pumped Storage Power Project produced electricity for the first time, and could generate up to 1,040 MW if needed to balance the load; the power traveled from the switchyard to substations in New Scotland, Fraser, and later to Leeds, New York, along three 345,000-volt lines. On the opposite side of the lower reservoir, the Power Authority also created the Mine Kill State Park, adding three enormous swimming pools, a bathhouse, playing fields, hiking trails, picnic areas, and parking lots. The historic Lansing Manor is on the property as well; built in 1819, its barn was neatly transformed into an NYPA visitor's center, and later the Manor was restored and refurbished by NYPA as a museum.

The construction of the Power Authority's first nuclear power plant, on the south shore of Lake Ontario near Oswego, did not involve the kind of decade-long delays and multi-billion-dollar cost overruns that became commonplace with nuclear plants just a few years later. In contrast to later runaway projects, wrote one historian, "the five-year construction period and the half-billion-dollar cost of the FitzPatrick plant were a reminder of the time when the atom had promised a cheap and plentiful energy future." While nuclear power was already regarded with concern by many long before the accidents at Three Mile Island, Chernobyl, and Fukushima, the number of nuclear plants in operation continued to grow for a few more years before falling off precipitously. Today there are 104 US nuclear power plants and 437 worldwide in thirty-one countries, according to the International Atomic Energy Agency.

The James A. FitzPatrick Nuclear Power Plant, as NYPA's trustees named it, was built adjoining the site of one existing Niagara Mohawk Power Corporation nuclear plant that had been operating since 1969, with another under construction. When completion of their second unit was delayed (for almost twenty years), the Power Authority contracted with Niagara Mohawk to appropriate some major components for use in NYPA's plant. The plant produced its first controlled nuclear reaction in February 1975, and five months later began commercial operation. NYPA was no longer exclusively in the business of hydropower.

Regulation Difficulties

While the need for more power grew through the 1970s, there was no guarantee that Con Edison and the other private utilities would or could stay ahead of the demand by building new plants, for which siting, financing, licensing, and construction had become more difficult. Part of the solution, provided for in legislation signed by Governor Rockefeller in 1972, was for the Power Authority to build the plants necessary to power the downstate subways and commuter rail lines.

As well, the following year NYPA applied to build a 765-kilovolt transmission line that would bring additional hydropower from Québec to a substation at Massena, near the St. Lawrence Project, and on to a substation at Marcy in Central New York, a total of 155 miles. From there, the power would travel downstate along existing lines. Since this would be the first line of such high voltage in the state (though not in the United States), lengthy regulatory delays were expected, and soon materialized, along with passionate opposition from farmers and other residents along its path. The Public Service Commission held extensive public hearings that addressed not only the route of the line and its environmental ramifications, but also the issue of possible effects upon the health of residents nearby and the threat of electrocution. Construction of the line was then approved. The Canadian power would prove invaluable for many years and in time NYPA transmission connections would extend all the way to Long Island.

But the Power Authority's plan for a nuclear plant in Greene County was ultimately canceled because of regulatory delays, which invariably increased costs. Then, in 1982, a coal- and refuse-fueled project on Staten Island was stopped by Governor Mario Cuomo, responding to intense and protracted opposition from the local press and politicians. Later, a pumped-storage project at Prattsville was canceled.

Besides, with the additional power from Québec, it seemed that demand could be met without additional generation by the Power Authority, which, including its Québec purchases, was supplying about 28 percent of the state's electricity at this point, benefiting almost all of the state's residences and businesses. Without question, this expansion was owed directly to FitzPatrick's fourteen years of leadership. "More than any other individual," wrote one historian, "he embodied the Authority's transformation from regional hydro operator to statewide utility."

He was also highly outspoken about the threat, in New York and across the United States, that protracted regulatory proceedings and civic opposition to new generation and transmission projects could damage the country's economic well-being. For this reason and others, FitzPatrick was deeply concerned about the public's indifference to the issue of power and the growing dependence upon foreign oil. In December 1976, he warned the state Assembly that the intent of the legislature was being thwarted by delays in the licensing procedures. As a result, not a single new plant had been approved since 1972, when, ironically, a bill was passed to accelerate the licensing process. A few months later, in his last major speech as NYPA's chairman, FitzPatrick drove home the notion that "there is no surer way to widen the schism between the haves and the have-nots in this country than to permit a situation in which electricity is either too expensive or not available in sufficient quantities."

Passing the Baton to Clark

Frederick R. Clark became FitzPatrick's successor in June 1977 and remained chairman for the next two years. An Albany banking executive and former state tax commissioner, Clark had a rude introduction to the challenges of providing energy: in July, New York City and Westchester were stricken by a second blackout. But the rest of the state was spared and once again, the Power Authority's operations were mostly unaffected, though the Indian Point 3 nuclear plant was shut down. This time, a series of lightning strikes at a Con Ed substation and along transmission lines near the Indian Point nuclear plants had tripped circuit breakers across metropolitan New York City. Simple human errors like verbal misunderstandings had exacerbated the situation, overloading and overheating transmission lines, which then sagged onto branches and shorted out.

As in 1965, the airports were closed and some four thousand people had to find their way through pitch-black subway tunnels. Shea Stadium went dark at the bottom of the sixth inning. This time, the civic bonhomie that had prevailed during the blackout twelve years earlier was hard to find. Looting broke out "in poor neighborhoods," according to the *New York Times*, and the Fire Department reported 1,037 fires throughout the city. About 1,600 stores were looted, so many of which were electronics stores that a major musical genre, hip-hop, evolved from all the expensive disk jockey equipment stolen that night. "The looters were looting the looters," one witness reported. Power was not restored until late the next day, and the overall economic losses were considerable. The message hammered home yet again was that a reliable grid had become indispensable to modern civilization.

Clark's role at NYPA was to complete projects that FitzPatrick had launched. Among his principal achievements was shepherding the 765-kilovolt transmission line through the final regulatory process and completing its construction. By December 1978, the 155-mile line from Québec to Massena to Marcy was fully energized, bringing a rich new source of power that would, in a few more years, stretch all the way to Montauk. To make full use of the line, a 20-year agreement was signed with Hydro-Québec to purchase 800 MW of "firm," or assured, power during the peak "summer" season of April through October for resale at no profit, principally to Con Ed and also to Rochester Gas and Electric for use by their customers. A separate agreement was signed to purchase "non-firm" energy from Hydro-Québec for use throughout the state at all times of the year. Within a year, these Canadian contracts had replaced the need for eight million barrels of oil.

Also during Clark's tenure, the start of service to Green Island in Albany County meant that the Power Authority was for the first time supplying all of the fifty municipal systems and rural cooperatives operating in the state. With advances in transmission, it became possible under Clark's leadership to provide assured power for the first time to the municipal systems serving Freeport, Greenport, and Rockville Centre on Long Island. Finally, power was reserved for the Town of Massena, which ultimately established its own municipal system after overcoming bitter resistance from Niagara Mohawk. But perhaps Clark's most enduring contribution was developing the new energy control center in Marcy, near the center of the state, reflecting the Power Authority's broader reach. Initiated under FitzPatrick to monitor and supervise all of NYPA's generation and transmission, the Frederick R. Clark Energy Center, as it was later named, became operational in 1980, replacing the Production Control Center at Niagara that had served the same purpose since 1970. The new center also included a transmission line maintenance and training facility. Now, with statewide generation from six plants and more in the planning stages, along with an expanded transmission network, the new location would prove to be the right choice.

Clark also launched a feasibility study, together with New York City's Mayor Ed Koch, for creating two small hydropower facilities at city-owned reservoirs at Ashokan in Ulster County and Kensico in Westchester. Thanks to their study, and to Clark's plan for another generating facility at the Hinckley Reservoir in Oneida County, first power was generated at the three sites in 1982, 1983, and 1986 respectively. These projects were followed by two hydro plants at the Crescent and Vischer Ferry dams along the Mohawk River. Altogether, these small plants provided an additional 13,000 kW of clean energy.

Clark left office in August 1979, by which time NYPA had evolved dramatically since FitzPatrick had taken the helm sixteen years earlier, when only the St. Lawrence and Niagara plants were operating. Much more than just the amount of generation had grown in that time: so had the value and significance of the Power Authority. Instead of exclusively benefiting, as it had at first, upstate residents and businesses, NYPA was now an indispensable supplier of power throughout New York.

CHAPTER NINE

THE DYSON ERA

Environmental Impact

On Christmas Eve, 1968, the *Apollo 8* astronauts became the first to orbit the moon and photograph the Earth rising above the lunar horizon. That image of our planet floating in space—the first global selfie, if one must—filled the cover of the environment-oriented *Whole Earth Catalog* at just about the time that the impact of man-made gases upon the atmosphere and the climate began to stir in the public awareness.

This impact, though recognized by scientists for decades, was still poorly understood. By 1908 the level of atmospheric carbon dioxide had risen from the centuries-old plateau of 278 parts per million (ppm) to 300. That year the Swedish scientist Svante Arrhenius wrote that "the enormous combustion of coal by our industrial establishments suffices to increase the percentage of carbon dioxide in the air to a perceptible degree." Thirty years after Arrhenius's work, Englishman Guy Stewart Callender confirmed that the average temperature of the planet's land masses had risen perceptibly since the dawn of the Industrial Revolution.

By the time astronauts orbited the moon in 1968, atmospheric CO_2 on Earth had risen to 325 ppm, and the issue of atmospheric degradation was merging with a wide array of environmental concerns that gripped the public imagination, starting with the threat of overpopulation. That same year, in his bestselling *The Population Bomb*, Paul R. Ehrlich observed that "the greenhouse effect is being enhanced now by the greatly increased level of carbon dioxide.... At the moment we cannot predict what the overall climatic results will be of our using the atmosphere as a garbage dump." Since that was written, the human population has doubled and continues to grow at a rate that is unsustainable and self-destructive.

In 1970, when the first Earth Day was observed globally, Congress created the Environmental Protection Agency and drastically amended the earlier Clean Air Act, giving it real teeth. As the use of coal by power utilities received fresh scrutiny, the solution appeared to be nuclear energy, and nuclear reactors began appearing across the country. The meltdown of a nuclear core did not appear to be a significant threat, nor was the disposal of nuclear waste a special challenge. Between 1970 and 1973, 114 new reactors were ordered, including Con Edison's Indian Point 2 and 3 in New York, which were intended to supplant the failing Indian Point 1.

But it was an event thousands of miles away that almost overnight focused the world's attention on all the issues of energy production.

OPEC Oil Embargo

On Yom Kippur 1973, the nation of Israel was surprised by a military attack from Egypt and Syria. The accelerated resupply of Israel's weaponry by the United States produced a ceasefire that halted the hostilities, but in response the Arab nations within the Organization of Petroleum Exporting Countries launched the OPEC (or "Arab") Oil Embargo, banning petroleum exports to the United States and other Israeli allies, and significantly reducing their own oil production. The embargo was originally proposed by Saudi Oil Minister Ahmed Yamani, whom the US press soon dubbed "Yamani or ya life."

This first "oil shock" profoundly readjusted the geopolitical landscape and transformed US energy use. Within six months, the price of oil quadrupled, from $3 a barrel to $12, unleashing recessionary forces throughout the US economy and provoking a profound reconsideration of dependence upon foreign oil. According to the State Department's Office of the Historian, the Oil Embargo also triggered efforts that "focused on energy conservation and development of domestic energy resources, [including] the creation of the Strategic Petroleum Reserve, a national 55-mile-per-hour speed limit ... and imposition of fuel economy standards" on all US-built automobiles.

The embargo caused an unprecedented jolt in the power industry as well. In fact, "all hell broke loose," wrote energy historians Richard Rudolph and Scott Ridley. "The electric power industry was dependent on oil for seventeen percent of its generation nationwide. But in some regions, such as the Northeast, generating plants were as much as sixty percent oil-fired.... Oil prices rose eightfold, and electricity prices went up nearly fifty percent."

In New York, which depended more upon foreign oil than any other state, the impact was nearly catastrophic by 1974. Forty-five percent of the state's electricity came from oil-fired generation at that time; a quarter-century later, that figure would be reduced to 13 percent. No less than 60 percent of metropolitan New York City's electricity was derived from oil, and Con Edison alone depended upon oil for three-quarters of its power production; so, although a sizeable rate hike was allowed by the Public Service Commission, this was the first time since Thomas Edison established the company in 1885 that its common stock paid no dividends, a fact that "hit the industry with the impact of a wrecking ball," according to energy analyst Leonard S. Hyman. "By September, prices for utility stocks had fallen by thirty-six percent."

The Power Authority had evolved and matured considerably under FitzPatrick, and though none of its own facilities burned oil, NYPA nonetheless had a de facto responsibility for the overall energy economy of New York. As an instrumentality of the state, it was obliged to help Con Edison. The investor-owned utilities had always regarded NYPA as competition, but the Power Authority was committed to collaboration on behalf of New Yorkers. It was only natural that NYPA promptly asked its municipal systems and rural cooperative customers to reduce their voltage use by 3 percent in concert with statewide voltage reduction measures to save power.

But this financial crisis also led to the Power Authority's purchase of two Con Edison plants that were under construction: the oil-fired Astoria 6, in the borough of Queens in New York City, and the nuclear-powered Indian Point 3, beside the Hudson River in the Village of Buchanan, Westchester County. The chairman of Con Edison, Charles F. Luce, had previously served as the administrator at Bonneville, where he had developed a sophisticated understanding of the role of public power. It was he who approached the Power Authority with a plan.

NYPA agreed to the purchase of the two plants despite some trepidations. To begin with, it meant that one of NYPA's facilities would hereafter depend upon oil and produce fumes and gases within New York City, and the new facility would require the Authority to take another bold leap into nuclear technology. But for Con Edison to declare bankruptcy could have unthinkable ramifications for the state; so, to pull the state's largest utility out of its financial crisis, Governor Malcolm Wilson ordered NYPA to purchase the Astoria oil-fired plant, along with Indian Point 3. After completing construction of the two plants, the Power Authority would sell most of the new electricity to public customers, as required by federal law for facilities financed with tax-exempt bonds. The seventy-six government entities being served by the end of 1977 included the government of New York City itself, its Housing Authority (NYCHA), the Metropolitan Transportation Authority, and the Port Authority of New York and New Jersey, as well as Westchester County and most of its municipalities, school districts, and other government entities. A portion of the power was assigned to Con Edison for its customers. So NYPA's resources were called upon to save a company that had fought with all its might against the creation of the Power Authority and delayed the St. Lawrence and Niagara power plants for decades.

Three Mile Island

The second oil shock began in January 1979 with the revolutionary turmoil in Iran, where some 37,000 workers went on strike at the government-owned oil refineries, aiming to depose Shah Reza Pahlavi. As skilled foreign workers fled the country, production was reduced by 4.5 million barrels a day. Other members of the OPEC cartel, led by Saudi Arabia, increased production, which left a shortfall of only 4 percent. Nonetheless, just the memory of the long lines at gas stations six years earlier produced *new* lines at gas stations, where cars often waited for hours. Most US vehicles of the era, when idling, burned more than a half gallon of fuel per hour, so an additional 150,000 barrels a day were wasted just waiting in long lines. President Jimmy Carter declared that this crisis was "the moral equivalent of war" and asserted that interference with tankers from the Persian Gulf would be taken as an attack upon US vital interests. His speech had little or no impact, and by the end of the year the international market price of oil had risen to $39.50 a barrel.

One morning, as this fresh shock reverberated through the economy, another crisis cast an even greater pall. On March 28, 1979, the Three Mile Island Nuclear Generating Station in Pennsylvania experienced the most dangerous commercial reactor accident in US history at one of its two units, TMI-2. Because of a failure in the cooling system, the unit suffered a partial meltdown and began leaking radioactive krypton-85 gas into the air, while 32,000 gallons of contaminated water were discharged from the Metropolitan Edison reactor. Although the threat was contained within three weeks and no deaths or injuries resulted, the weaknesses that the accident exposed provided a blistering experience for the nation as a whole, and debunked the wishful thought of nuclear power as *the* energy solution.

The problem in the TMI-2 cooling system—a stuck valve—might have been easily resolved, but was instead gravely compounded by human error: the poorly trained operators misinterpreted events and warning lights, and then took the wrong remedial actions. There were other critical valves that were closed at the time for "routine maintenance" in violation of Nuclear Regulatory Commission rules. Pennsylvania's Lieutenant Governor, William Scranton III, who initially was misinformed by Metropolitan Edison, issued a report reassuring the public; but the next day Governor Dick Thornburgh called for a voluntary evacuation of pregnant women and pre-school-age children residing within 20 miles of the 906 MW plant. That, of course, produced near-panic among many of the 663,500 residents within that radius, and when a hydrogen bubble threatened to explode in the pressure vessel, more than 140,000 people left the area. Plans for an orderly evacuation did not exist.

Twelve days before the accident at Three Mile Island, a film called *The China Syndrome* had opened, in which an accident at a California reactor threatened a nuclear meltdown that could theoretically burn through the Earth's core to China. The film, starring Michael Douglas, Jack Lemmon, and Jane Fonda, had initially drawn mixed reviews, and a spokesman for nuclear power had denounced it as "character assassination of an entire industry"; nonetheless, after the accident at TMI-2, the film became a box-office hit. From this fictional accident, audiences learned that some reactors *and* nuclear waste facilities are always vulnerable to seismic shocks that could, in a worst-case scenario, contaminate swaths of the country for millennia to come.

From the real-world crisis, Americans learned that the nuclear industry was capable of being complacent, cavalier, dishonest, and utterly unprepared for a serious mishap; in fact, barely competent to operate facilities. And they learned that the Nuclear Regulatory Commission (NRC), created four years earlier, was, like its predecessor the Atomic Energy Commission, far too closely wrapped up in the industry to serve as its watchdog, even for critical issues like safety. Seven years later, after the much more serious nuclear catastrophe at Chernobyl, a congressional committee found that the NRC was still enthralled by its "Coziness with [the] Industry" and failed to "maintain an arm's-length relationship" with it.

Naturally, New Yorkers began to wonder what they would do if an accident or earthquake disrupted the spent-fuel pools and Units 2 and 3 at

Indian Point, operated by Con Edison and the Power Authority respectively. As well, six years earlier the Long Island Lighting Company (LILCO) had begun the construction of a nuclear plant in Shoreham, Long Island; now, for the first time, many nearby residents questioned how they would evacuate in the event of an accident there. No answers were forthcoming. The Pennsylvania accident had revealed a slew of failures and a dearth of training, foresight, planning, and forthrightness in the commercial nuclear industry, and demonstrated that nuclear power was not a painless panacea for the energy crisis that now challenged Americans' lifestyles. At Three Mile Island, the clean-up alone cost a billion dollars and took four years. The NRC and the nuclear industry as a whole took the accident as a wake-up call, responding with much tougher rules as well as periodic emergency response drills and exercises. It was less than a month later that NYPA's trustees voted to sell the assets for a proposed 1.2 MW nuclear plant in Greene County.

It was during the 1970s, then, that Americans began to recognize *all* the costs of energy. The magically transformative power of electricity had, until now, always outweighed its "inconveniences," going back to the first plant on Pearl Street, which required a never-ending parade of horse-drawn wagons filled with coal while spewing black smoke from its six locomotive engines. But by now the *full* price of that magic had become painfully apparent to most Americans: the cost of heating their homes and leaving appliances on all the time; the cost to their health of atmospheric lead from gasoline (until the mid-1970s), as well as smog (particulate pollution), largely from coal-fired power plants, which has a medically measurable impact when cases of asthma are monitored; the cost of suburbanites commuting, and the trucking of fresh produce from California's Central Valley; the cost of a nuclear accident, even without widespread, long-term devastation; the cost of storing nuclear waste safely for centuries to come; the atmospheric cost of leveling vast dominions of forests that serve as the planet's "lungs"; the geopolitical cost of a foreign policy focused upon petroleum imports; and finally, the looming costs of an unstable climate for as long as mankind continues to disrupt it and beyond. All these problems associated with energy were obvious by 1979. The solutions were not.

The Dyson Makeover

That year John S. Dyson was elected Chairman of the Power Authority. At the age of 36, he was a Cornell-educated New Yorker from Dutchess County who had already served as State Commissioner of Agriculture and Markets before becoming Commerce Commissioner, where he is best remembered for developing the "I Love New York" program, an effective and enduring campaign to promote tourism. The *New York Times* once characterized Dyson as "a man who never wanted to be a faceless bureaucrat; a commissioner with ... a penchant for the flamboyant gesture or outrageous quote."

Now Dyson was preparing to take on a national crisis, with a surplus of self-confidence and a multifaceted effort "to break OPEC's stranglehold on this country." As he declared in a speech to the cadets at West Point, "The United States must regain control of its destiny by reestablishing its energy independence." The oil cartel's grip on the world *could* be overcome, he explained. "All we have to do is to get the demand under the supply and all the OPEC countries will start fighting with each other, and then we can break the prices. If we're going to break the stranglehold of OPEC, it's got to be done in New York. We can break it here ourselves."

With the assent of Governor Hugh L. Carey, Dyson announced a crusade to reduce by half the state's use of oil within a decade. His Ten-Point Plan included adapting the Astoria 6 plant (later renamed the Poletti plant) to dual-fuel generation so that it could burn natural gas in the summer months when home heating did not require it. Emphasizing *clean* energy generation, he also proposed that New York should purchase more hydropower from Canada and develop the kind of small hydroelectric projects that his predecessor, Clark, had explored. His other emphasis was upon aggressive energy conservation initiatives to be implemented by NYPA and other state agencies. Although the New York legislature proved less than helpful in some areas, and regulatory problems scotched some of Dyson's proposals, others would have a significant impact.

When asked how he ranked his achievements at the Power Authority, Dyson replied without hesitation. "First, I would say, was putting the Authority on one financial footing. When I got there, every project had its own finances and had its own bond issue. And what we did was to take the great strength of Niagara Falls and put it behind everything, by reissuing a whole new set of bonds ... and what they call technically defusing the old bonds with the leadership of Goldman Sachs and Salomon Bros. and others, to create one Authority with one set of books and one set of bonds. That prepared us for a big project like [the transmission line] Marcy-South—a $500 million project, which had some financial risks—so that we could issue the bonds for that backed up by the great strength of the rest of the Authority's finances. [With that in place] we planned a line under the Long Island Sound, and then after me, my fellow board member, colleague, and friend, Dick Flynn, went on to complete it, along with a new plant on Long Island.... Marcy-South became the great backbone for allowing Canadian power in."

Others gave Dyson high ratings for a broader achievement. James Cunningham came to the Power Authority at the same time as Dyson became Senior Vice-President for Public and Government Affairs and remained for fifteen years, often personally responsible for much of the heavy lifting needed to advance the projects launched by Dyson and his successor, Richard Flynn. Cunningham watched as Dyson, with his help, "changed the morale and the image of the Power Authority. "When we got there," said Cunningham, "we were losing people left and right. He describes how, in their first few months, they were saying goodbye to slews of retiring engineers and administrators week after week, until Dyson asked, 'Jesus, Jim, why are *we* staying?'"

"We made quite an effort at professionalizing the agency," Dyson said recently. "One of the key parts of it was that we undertook an employment study of comparative salaries with the Hay Associates group... and found that we were way behind [other power companies]. I proceeded, with my colleagues on the board, over time, to raise those salaries, with Governor Carey's permission, at a higher rate than [that of] the other state employees: we were competing in an electrical industry, while they were competing in a government industry. We then had two nuclear plants and we had to have people who were at the top of their trade and we had to pay them comparable salaries. That led to retaining a whole generation of young people seeing that working for the Power Authority would be as remunerative as one of the private utilities down the street, and in some ways this was much more fulfilling since we were moving, we were *doing* things, we were doing small hydro, we were converting plants from oil to gas, fixing up nuclear plants, changing steam generators, and the lines under the [Long Island] Sound. When you think of everything that was going on, you wouldn't see all that at Con Ed in a whole career."

As for changing the image of the organization, Dyson began by reducing the formal name chosen by FDR in 1931, the "Power Authority of the State of New York" to the "New York Power Authority," except in government and finance documents. A new logo was designed, and a reinvigorated effort at outreach to the company's stakeholders began with a revamping and expansion of what would thereafter become the Communications Department.

Conservation and Efficiency

An important and enduring legacy of Dyson's NYPA chairmanship was his emphasis upon a little-used approach to increasing the amount of electricity available by improving the end-users' energy efficiency. "Little-used" because, of course, the private utilities up to this time sought to sell as much power as possible. In retrospect, it is remarkable how much energy was being squandered by American homes, businesses, factories, and government facilities ... and how much is still wasted today. Dyson was among the first government officials to address energy conservation comprehensively, exploring all the ways in which the prodigious waste of electricity could be reduced. He was imaginative and unafraid to undertake the very specific, detailed efforts that would be required.

A substantial proportion of this electrical waste came from heating, ventilation, and air conditioning (HVAC) during New York's frozen winters and blistering summers. Much of this waste resulted from inadequate insulation and antiquated equipment in homes, most of which were served by investor-owned utilities, not by NYPA. How, then, could Dyson organize and finance an effective effort to conserve electricity across the state? The answer, he concluded, was for the Power Authority to teach conservation by example wherever possible, and thus provide a yardstick to demonstrate how New Yorkers could save money through energy efficiency. Just as governors Hughes and Roosevelt had envisaged, NYPA would set a higher standard for all power companies, in this case by encouraging and helping customers to use *less* electricity.

After six months in office, Dyson introduced the Button Up program for the 125,000 residential customers served by the state's municipal electric systems and rural electric cooperatives, offering "free home energy surveys [that] would allow consumers to save energy dollars while helping to break the stranglehold of OPEC on the state and nation." Building upon a pilot program NYPA had tested in 1974, Dyson's goal, he explained, was to "show homeowners, through this program, how to button up their homes, as well as their wallets and pocketbooks." Dyson wanted this to be "the most thorough attempt made to date to encourage home energy conservation," and "to put the Power Authority and the municipal and rural cooperative systems of New York in a leadership position with these efforts." The fifty municipal and co-op communities, all of whom were Power Authority customers, had pledged their support to the program.

More than 100,000 residential customers in these communities received home energy surveys to fill out, along with detailed instructions. The surveys were returned to the Power Authority for computer analysis, after which the homeowners were provided with custom-tailored energy conservation recommendations, including details on costs and savings. "Basically," Dyson explained, "our intent is to put all needed information in the hands of residents so they can make informed decisions about saving energy and money." NYPA would also learn more about the customers' power usage, information that could be shared with the private utilities and their customers; because the first step in improving energy efficiency is to identify where it is needed.

This do-it-yourself approach to energy audits meant that no one from the power company was going to come inside your home. But in a separate, experimental part of the program, NYPA offered 25,000 Long Island residents free home inspections conducted by conservation experts at the residents' convenience, weekdays or weekends, to check from cellar to attic for areas where significant energy savings could be achieved.

The most distinctive aspect of the Button Up program was that it offered low-interest loans to homeowners to pay for insulation or replace old heating equipment. Dyson called this one of the most aggressive campaigns ever conducted in the area of energy conservation, which included extensive advertising, direct mail, and seminars. And unlike other companies' home audit programs, this was free. The program's ambition was to save almost four million gallons of oil in two to three years among these fifty communities alone. If carried out statewide, such efforts might save as much as 200 million gallons of oil and $200 million (in 1980 dollars) of costs for consumers every year.

Dyson also foresaw a broader conservation program being developed for industrial customers. He noted that private industry had "the best record of any sector in our society in saving energy, because of their recognition of the profit motive to do so," but "those industries enjoying the benefit of our valuable low-cost power" must nonetheless commit themselves to energy conservation. The same must apply to public agencies receiving Power Authority electricity. Button Up and similar statewide programs, said Dyson, "can make New York the national leader in proving the American people have the willpower to mount a do-it-yourself fight against OPEC oil." The idea that you could strike a blow for your country by sealing cracks in your window frame added a patriotic luster to household drudgery. It was also a perfect illustration of the new environmental catch phrase printed on a million bumper stickers: "Think globally, act locally." That motto remains at the core of NYPA's guiding principles, incarnated today in the word "sustainability."

Economic Development

The Authority also took new steps toward economic development, in which NYPA has played a much larger role ever since Dyson's leadership. That was because he saw the Authority's low-cost power as a valuable tool for motivating companies to create and retain jobs. "I inherited," Dyson explained, "as a matter of timing, the expiration of many of the original contracts that Moses had signed with Alcoa and Reynolds, and all kinds of chemical companies like DuPont and others in Niagara and also St. Lawrence. It seemed to me an opportunity to make the point that the Power Authority, as one of its missions, provides this low-cost electricity, so we ought to figure out better ways to connect the employment with electricity."

The foundation for economic development is the retention of existing jobs. Alcoa, the Aluminum Company of America, had been operating continuously in Massena since 1903. Together with Reynolds Metals (they merged in 2000), which had started out producing tin foil for Reynolds Tobacco cigarette packs, Alcoa was NYPA's first customer when the St. Lawrence plant powered up in 1958 and a mainstay of the economy in Northern New York. The two companies' original long-term contracts were due to expire in the 1990s. Dyson negotiated with both companies and in August 1981 signed new contracts at a ceremony at the St. Lawrence project for the continued supply of St. Lawrence hydropower to the companies' Massena plants through 2013.

The date for the two signings was chosen to emphasize the long-standing ties between the Authority and the aluminum industry, because the next day celebrations were held marking the 50th Anniversary of the Power Authority Act with which Governor Franklin D. Roosevelt had created the state's own electrical provider (notwithstanding the twenty-seven years before it actually provided electricity). NYPA marked the event by renaming the plant the St. Lawrence–Franklin D. Roosevelt Power Project, and featured the late president's son, Franklin D. Roosevelt, Jr., in the celebration. Later that year, NYPA commemorated the anniversary with a special program at the Hayden Planetarium in New York City.

Dyson also launched the "Juice for Jobs" initiative, whereby the state offered low-cost NYPA electricity to those industries and businesses that made a commitment to maintain or increase the number of their employees. Under this pioneering program, NYPA helped assure the creation or retention of some 15,000 jobs. Thirty years later, programs that Andrew Cuomo has created like ReChargeNY are direct descendants of Dyson's efforts. In fact, in 2014 the Power Authority amended its mission statement to read: "Power the economic growth and competitiveness of New York State by providing customers with low-cost, clean, reliable power and the innovative energy infrastructure and services they value."

For NYPA, lowering the cost of power entails minimizing its own operating costs, including those in its office space. One step that Dyson took was to start moving the Authority's main offices out of Columbus Circle in Manhattan and relocating most employees to a building on Broadway. Then, after careful deliberation, he chose the city of White Plains, Westchester, for a new corporate location, a sensible choice that is close enough to the City and closer to Indian Point. It also proved to be a better location for recruiting employees. And not by accident, it was a shorter commute for several executives who lived in the northern suburbs. At first, NYPA relocated 20 percent of its New York office staff, about 150 employees, mostly from the Nuclear Generation Department, to White Plains, which became the Authority's main offices after it purchased the building in the 1990s.

Exploring Renewable Resources

By 1981, NYPA sold more electricity (29.7 billion kWh) than any other federal public power organization, but thanks to its exceptionally low rates, revenues for the year ($974.1 million) only ranked third. Dyson was still considering every possible method of generating electricity to reduce the state's dependence upon oil and to meet the unceasing growth in demand, which was greatest in New York City, of course. Given the mounting concern about anthropogenic climate change, he considered the options for renewable resources. Although the cost of building an industrial wind farm back then, and the complex regulatory issues involved, made it a poor gamble, NYPA nonetheless cosponsored a modest, 10-kilowatt wind energy system at Morrisville College to explore the future potential for wind power projects as an alternative to fossil fuels.

Hydropower was the best option, but time was of the essence, given the threat of another oil embargo, possibly more severe. There were few natural locations for water power near New York City, and more distant options would require major transmission projects that were sure to involve eminent-domain conflicts with property owners along the path, as would prove to be the case with newly proposed high-power transmission lines from Canada.

Dyson considered the feasibility study conducted by his predecessor, Frederick R. Clark, and New York City Mayor Edward I. Koch, who had proposed generating power along the system of reservoirs and aqueducts that provided most of New York City's tap water, some 550 million gallons a day. Built in 1915, the series of 19 reservoirs and controlled lakes begins a hundred miles to the north in the Catskill Mountains; from there, fresh water flows by gravity at a rate of 4 feet per second to some 70 communities bordering the waterways. It travels along the Catskill Aqueduct and through a tunnel 1,000 feet beneath the Hudson River to the Kensico Reservoir in Westchester County, where it connects to New York City's distribution system. Clark's feasibility study had established that the quality of drinking water would not be affected by passing through turbines.

So in his first year, Dyson launched several small hydropower projects at existing reservoirs where the dams were well-suited for generating facilities. The first of these plants began operation in November 1982 at the Ashokan Reservoir west of the Hudson River, some ninety miles from New York City. The plant's capacity was just a tiny fraction of Niagara's (4,750 kilowatts compared to 2,400 megawatts), but it connected easily to the grid without hundreds of miles of new transmission lines and it did not affect the atmosphere. The same was true of NYPA's second small hydropower unit at Kensico in Westchester, which first produced power in 1983. These units generated the first new baseload hydropower for NYPA since Niagara had begun generating in 1961. A project at the Hinckley Reservoir, just north of Utica, was next; it began generating 9,000 kW in 1986, when it was renamed the Gregory B. Jarvis Plant to honor the astronaut, a local resident, who had died on the *Challenger Space Shuttle* that year. Dyson also applied for federal licenses to expand two small plants on the Mohawk River, at Vischer Ferry and Crescent, which, when fully operational, added about 4,000 kW to each.

Inexpensive hydropower was available for purchase from Canada. Within his first year, Dyson began seeking approval for the construction of two 345-kilovolt transmission lines between NYPA's Niagara Power Project and Ontario Hydro's Sir Adam Beck complex. This would allow the two companies the flexibility to exchange support when it was needed—and to avoid a repeat of the Northeast Blackout of 1965 that had impacted both countries.

Also on the St. Lawrence River, further east, across from Massena, was NYPA's partner Hydro-Québec, the world's largest hydroelectric producer. Its sixty generating facilities throughout the province, wholly owned by the government of Québec, produced a surplus of power, and Québec officials were eager to sell low-cost electricity. In 1978, the Power Authority completed construction of its 765-kilovolt transmission line from Québec to Massena and 134 miles on to Marcy, near the Frederick R. Clark Energy Center (CEC), which became operational under Dyson. The CEC serves as NYPA's hub, coordinating the flow of electricity from all its plants and dispatching power where it is called for, as the total load changes constantly. Today, the CEC exchanges information every six seconds with the New York Independent System Operator (NYISO), which oversees the entire state grid and supervises the minute-to-minute electronic auction for power generated by private power companies and NYPA.

The additional Canadian power flowing to Marcy was a tremendous boost. After just two years, Dyson announced, New York had saved more than a billion gallons of oil and about $100 million by purchasing renewable hydroelectric power from Québec—and there were more savings to be had where that came from. After lengthy negotiations, Dyson and his Québec counterpart signed a contract for the sale of more than 100 billion kWh between 1984 and 1997.

But that additional power would require another transmission line, named Marcy-South, to serve, along with NYPA's customers, those of Con Edison in New York City and Westchester as well as LILCO, and Orange and Rockland Utilities. Given the elevated price of oil, NYPA estimated that consumers would save $186 million a year with the Canadian purchases. But the investor-owned utilities were dragging their feet about the transmission line, Dyson recounts, "so [NYPA President and Chief Operating Officer] George Berry and I went to the meeting of the Power Pool and I said to them, 'We are very close to a deal with Québec hydropower, and we're going to need this line. You guys are always accusing me of being a socialist, so I'm giving you the first chance. You can build this line, and the Power Authority will put in its usual 11 percent of the Power Pool; or we will build the line and then you don't have to put anything in. So George and I are going to go down the hall to get a cup of coffee and we'll be back in thirty minutes, then you tell me whether you want to have the Power Pool build it or you want us to build it. But I've been instructed by the Governor to get on with this.' So we walked down the hall and had a cup of coffee, and when we came back, they said, 'We want you to do it.' George and I thought the votes were not there [for the utilities to build it], because Con Edison and Niagara Mohawk didn't want to do it, and they had a lot of the votes."

So in 1982, the Authority applied to the state for permission to build a double-circuit, 345-kilovolt line from Marcy to East Fishkill in Dutchess County at an estimated cost of $509 million. It would require 1,532 towers, 3,270 concrete foundations, 2,132 miles of wire, and 261,000 insulators. The Public Service Commission held hearings in 1983 where enough passionate opposition was expressed by some landowners in the eight affected counties to delay the project and continue the hearings into a second year. Some of the opposition was focused upon the purported health threat of living below electromagnetic fields, while others objected to the "visual pollution" of the towers, which ranged from 80 to 148 feet in height.

"I led that fight," James Cunningham remembered. Discussing the challenges of the Marcy-South line, he noted that, "in the utility business, it's extremely difficult to get *anything* built, but at least with power plants the opposition is contained within that community.... A power line that runs five hundred miles through counties and towns, through so many backyards, affecting so many people... and so many political people. The Marcy-South line went through, I think, forty-four towns in eight counties." There were "one or two innovative things, in that struggle, which were later replicated by utilities across the country. I actually went around the country and interviewed builders of transmission lines.... Based on discussions I had with leaders of utilities that had horrible experiences building transmission lines, they told me what had worked best, so I cobbled those together and came up with the public approach to Marcy South.... For the team that got it built, it was a transformative process, and in fact utilities as far away as Japan came in to see how we got it done; we got calls from Australia."

"By September 1982," according to a NYPA publication, "ten large and stormy meetings in otherwise small and quiet towns had been held" before Cunningham changed the format. By then he had learned the hard way that it was counterproductive to start public hearings in an auditorium, where passionate opponents and hecklers immediately attracted the attention of the press. Instead, he learned to begin each public hearing at the entrance of the auditorium, providing the public an opportunity to ask questions and express their concerns at separate, clearly marked tables addressing each potential issue: financial, environmental, access through private properties, and health. (This last was because of concern in the 1980s about the purported threat of electromagnetic fields or EMF, which has been well debunked scientifically.) Residents were thereby informed about each of their concerns separately, as was the press. This approach defused the protesters' outrage, and by the time the general meeting and the Q&A period began inside the auditorium, the press had usually left. That simple strategy produced very different outcomes with scores of these hearings.

The new line would follow NYPA's existing rights-of-way for 70 percent of the distance, but in other areas it ventured into ecologically sensitive land. In 1985, after considering some 12,500 pages of testimony, the Public Service Commission finally overruled the objections and approved the project, subject to rerouting in several places and upgrades at substations. Construction began that July under Chairman Richard M. Flynn.

Nonetheless, opposition continued to swell. By 1988, a group of ninety-four landowners sued the Power Authority for degrading their property values, despite studies submitted by NYPA, "which have withstood cross-examination in many court cases," showing no difference in the price per acre in recent sales and "no persuasive evidence of increased risk ... associated with power lines." Residents nonetheless worried that their health would be affected by electromagnetic fields, and argued that the line was unnecessary because extra power could still be obtained by energy conservation. NYPA also granted every town *and* county $55,000 for each mile the line crossed within its borders, without any conditions on where the money was spent. Nevertheless, Dutchess, Orange, Otsego, and Sullivan counties joined forces and funds to fight it. An editorial in the *Times Herald-Record* supported the need for Marcy-South but admitted the line would be "a mixed blessing." Construction proceeded, and in May 1988, after Dyson had left, the line was energized. It had cost an additional $161 million but saved a similar amount every year or two.

Dyson had his setbacks as well, the largest of which was his effort to create the Residential and Rural Energy Authority, a statewide public agency that could therefore qualify to buy Power Authority hydropower and then distribute it to individual consumers who were outside the existing municipal and cooperative systems. The aim of this new agency would be to provide equal shares of hydropower to all New Yorkers within a decade, under a "one man, one volt" concept. This proposal produced a roar of opposition, not only by the investor-owned utilities, which opposed the plan, but also, most intensely, by the municipal systems and cooperatives. Representatives of public power from around the country also expressed their opposition. "It would mean economic havoc for present recipients" in New York, wrote the Executive Director of the American Public Power Association in the *New York Times*, and "dilution of the value of Niagara's output to the point that individual consumers would realize negligible benefit." In 1982, the state's legislators, some of whom were as beholden to the private utilities as those who had opposed Governor Hughes seventy-five years earlier, simply refused to act on the bill that Dyson had drafted.

Then newly elected Governor Mario M. Cuomo created a commission, led by a former Power Authority Trustee, Robert I. Millonzi of Buffalo; and fifty-two municipal distribution agencies (MDAs) were created to qualify as NYPA customers. But once again the Legislature refused to act upon the proposal. In any event, federal courts decided that the MDAs would not qualify for public power since they would not own the distribution lines, but instead planned to lease them from the commercial utilities.

Still, with only a few stumbles, Dyson cast a giant shadow in his years leading the Power Authority, making energy efficiency a guiding principle for NYPA and demonstrating the path toward freedom from foreign oil. His success can also be measured by the growth of the Authority's role during his tenure, which in his last full year provided 36 percent of New York State's electricity, less than 2 percent of which was generated from burning oil.

CHAPTER TEN

FLYNN AND THE THIRD GENERATION

Expanding NYPA's Reach

"I had kind of a long involvement with NYPA, going back to my dad, Edward J. Flynn," explained Richard M. Flynn in a 2015 interview. "He was Governor Franklin Roosevelt's New York Secretary of State and actually certified the Power Authority Act in 1931. Roosevelt routinely consulted with him. When Roosevelt went to Yalta he took him with him on the boat, and sent him on to talk with the Pope, and then to Molotov about the fate of Catholics in Eastern Europe.... My mother and father had a close relationship with Eleanor and Franklin. My dad understood the type of person that Roosevelt was: he might have an ultimate objective but he might switch course to try to get there, and dad could intelligently follow all that and constructively help him achieve his objective."

Having grown up in that milieu, and after serving as a NYPA trustee for a decade, Richard Flynn was well versed in the tactics of political negotiation that he would conduct as Chairman and Chief Executive Officer of NYPA when he was elected by the Trustees in June 1985 to succeed John Dyson. But Flynn also brought to the job a surefooted understanding of rate structures. "I had a background in rates," he explained, "because I'd represented a couple companies in the cable television industry."

Flynn insisted that NYPA had to be truly competitive in its field, which was "always a strong guiding principle with me," he said. "The Power Authority, one of the largest non-federal public power producers, advanced the acquisition of a new source of energy on a *competitive* basis. It was a unique, competitive wholesaler. The private utility company would simply contract for the building of a plant and look for its sustenance in the rate of return that it was allowed, based on the finished construction costs. But the Power Authority really had no franchised territory (with the exception of the municipalities, and later the government customers in New York City and Westchester County), so to successfully build and market the power from any of its plants, it had to be darn sure that it did it for less and did it efficiently, one of the great disciplines with which most of us operated."

Flynn was echoing the point that Governor Hughes had hammered home a century earlier: that the investor-owned utilities did not care how much they spent building a new plant, because the rates they charged would ultimately be based upon that (often inflated and usually wasteful) cost, with profit added on. So, once again, according to Flynn, it fell to the Authority to provide an essential yardstick, even in the construction of new sources of electricity, with which the construction costs of other power producers could be compared.

"During my time and John Dyson's," Flynn went on, "the Authority was producing about a quarter of the electricity in New York State [28 percent; with purchases from Québec and other sources included, 36 percent] and most if not all of that was conceived, initiated, and sold on a competitive basis. All of NYPA's sources of energy had to find markets because it didn't have a franchise; so to sell the power it *had* to build it for less." What made this possible, Flynn said, was that, "at the end of the great dominance of the private utilities, they became very inefficient, which resulted in many badly managed and overpriced power facilities. We did our math with every project and made darn sure that power was going to be sold for less. We also focused, relatively speaking, on good, clean sources of power: our history bears that out."

He also demanded discipline from prospective contractors, who were accustomed to winning jobs with phony low bids only to report enormous cost overruns later. Flynn warned every company bidding that "if you make a bid, you better mean it, or you'll never get another contract with the Power Authority. That seemed to have an effect: most of the contracts I was involved in were finished ahead of schedule and under budget. And all these were broadening the Authority's range, aimed at bringing power downstate."

Bringing State Power to Long Island

The most urgent task at hand when Flynn took office in 1985 was the construction of the Marcy-South line, because the contract with Hydro-Québec could be canceled if the line was not completed before September 1988; the line was only near the end of its licensing phase. Jim Cunningham's ambitious public outreach effort had won approval for rights-of-way with the forty-four towns and scores of landowners directly affected, but the construction was challenging as well, covering all sorts of terrain and geology during every kind of weather, and passing under the Hudson River. Nonetheless, by May 1988 the 207-mile line was fully powered, and at a dedication ceremony the next month, Flynn rightly called this "a major accomplishment. Less than six years after we applied for permits, and less than three years after we started building, we have added a key new link to New York's electric transmission network."

But from Dutchess County there was still further to go in order to reach Long Island, where power was needed most. This project was, in effect, the final 27-mile leg of the Authority's transmission network from Canada. "We all realized that the transmission situation for Long Island was terrible," explained Flynn. "It had almost no access to clean, low-cost power. That was why we got into this 345-volt, underground, underwater connection under the Long Island Sound, completed during my leadership."

Construction began in May 1989, and it was completed two years later, said Flynn, "eighteen months ahead of schedule at a cost of $330 million, $46 million under budget. It was the first large, underwater cable contract for placing the cable on the bottom of the Sound rather than putting it in ducts." The Sound Cable Project, as it was called, was a real challenge for Jim Cunningham and his public hearings. The first eight miles, from Yonkers to New Rochelle, ran underground beneath towns and private properties across Westchester. Then the cable was laid along nine miles beneath the Long Island Sound by a ship leased from Pirelli Cable Corporation in Naples, Italy. The line reemerged at Hempstead Harbor and again traveled underground to a Long Island Lighting Company substation in Garden City. The low-cost power allowed for the creation of 700 new jobs and the retention of 2,300 more.

At the official ceremony upon completion of the project, Governor Mario Cuomo spoke with passion about the value of the Power Authority, evoking political theory from Abraham Lincoln.

> Government is the coming together of people to do for themselves collectively what could not be done, or done as well, individually. You couldn't have done this project just with the private sector. Public power is absolutely essential in this state, at this moment. You try laying this cable through Nassau and Westchester as a private combination without any government involvement. What private sector operation that you know has shown such extraordinary efficiency that they could bring in the job, not on time: this job is eighteen months early! Now when's the last time that happened? Your government did it, in the form of the Power Authority. Not a Democratic government, not a liberal government, not a conservative government, not an ideological government, but an intelligent government. Some things government does, it *must* do, because no one else will.... [In this state] we are very good at government, and the New York Power Authority is maybe one of our proudest boasts.

> You almost never hear about the New York Power Authority, but you should. It would do us all a lot of good: you can see government at its very best. You've done something here that they said couldn't be done.... They said, "You'll never get it past property owners, you'll never get past the lawsuits, you'll never get it down, nobody in the country makes that cable, you can't put it under the Sound, the environmentalists will kill you, what if there's an explosion, what if a fisherman using a big hook pierces it, what if it electrifies and kills the fisherman ... then it kills the fish! They'll never let you do it." Well, here you are. How did you do it? *Government did it*, and that's something to be proud of.
>
> The whole thing is absolutely saturated with intelligence and wisdom. You're replacing oil! You fight wars over oil in this country and everybody knows it. Eisenhower knew it, Nixon knew it, Ford knew it, Carter knew it, they all knew it, and they all told us, "For God's sake, stop the addiction! You will go to war over oil!" And we just did. [The Gulf War had ended three months earlier.] *And everybody knows it.* You can dress it up, you can pretend, you can make great speeches; *it was about oil.* In the Persian Gulf. And it will be again, unless you come off the addiction. We aren't off the addiction with this cable, with the hydropower it will bring down in 1995, with the other alternate energies, but we're moving in the right direction. And if someday the national government comes up with an energy policy that will synchronize with ours, well then, we will indeed have made tremendous progress.

Still, Long Island was going to need more electricity. LILCO knew that, and had a location for a small, 135,000-kilowatt plant in Holtsville. Flynn, always interested in advancing "the acquisition of a new source of energy on a competitive basis," entered a bid for NYPA to undertake the construction of a natural-gas-fired plant that was "combined-cycle"; that is, after the conventional turbines produced about 100,000 kilowatts, the exhaust heat would then be reused to spin a second turbine.

Flynn recalled that:

> In 1990 we won, in competitive bidding, [the contract to build] the gas-fueled power plant in Holtsville. LILCO had gotten more than twenty proposals to build it. How we did that was, it was a very efficient plant, and clean, gas-powered, and it was combined-cycle, which in those years was not widely utilized, so we used that as our model. And *we* got the gas. We went out to Western Canada—we were really doing it right!—and we got low-cost gas, and we got a company, I think it was Iroquois, to transport the gas across the country by a series of pipes. It took us a long time to do it but we got it done. They call it now the Flynn plant, though I had nothing to do with that, but I kind of like it. It epitomized a new direction for the Authority and a wholesome new direction for the utility industry as a whole.

Today, the Holtsville facility, owned and operated by NYPA, sells all its power for distribution by the Long Island Power Authority (LIPA).

Flynn extended many of Dyson's conservation plans. The new Watt Busters initiative built upon the success of the Button Up program, providing home energy audits and weatherization at no charge. Then in 1990 he added his own important contribution.

> We started with the city of New York an energy efficiency project that had significant impact through the decade and beyond. The High Efficiency Lighting Program (HELP), which was initially aimed at the Authority's governmental customers in New York City and then Westchester County. Lighting, in offices, all the city's museums, schools—this was all a target for a very efficient energy conservation program.... We financed it, and recovered its costs by sharing in the savings of energy bills, after which the customer retained all the savings. And not only for the city and Westchester, but it was expanded to cover all of the state government facilities in

> 1991, all the public schools on Long Island in 1992, public schools and community colleges statewide [in] '93, and municipal and county governments in 1994, after I left. The Authority received an award from FERC for the best program of its kind in the country: leadership in a new area that the Authority should most definitely have been involved in.

The Shoreham Nuclear Facility

On Flynn's watch, it also fell to the Power Authority to drive a final stake through the heart of the most extravagantly wasteful fiasco in the history of US power. In 1965, near the peak of the national infatuation with nuclear power, LILCO announced plans to build the first commercial nuclear power plant on Long Island, where demand for electricity was increasing by 10 percent annually; the project had the ardent support of elected officials. The following year LILCO purchased 455 acres in Shoreham, on the north shore of Long Island, and planned to construct a 540-MW boiling-water reactor at a cost of $65 to $75 million, to be completed and online by 1973.

By 1969 LILCO's mushrooming ambitions included building two more nuclear plants nearby and enhancing the planned Shoreham facility to 820 MW. These major changes delayed construction permits for years, giving the well-heeled opposition around Lloyd Harbor time to mount a fierce opposition. Meanwhile, the projected cost of construction continued to rise, approaching $2 billion by the late 1970s, largely because of "astonishingly low worker productivity." As well, the architect/engineering firm of Stone & Webster had negotiated a "cost-plus" contract, according to which the longer the project took, the more the company earned. That was how its original $16.6 million contract rose to $347 million. Community organizers continued to ramp up their opposition, which initially focused on the 400 to 500 pounds of radioactive waste that Shoreham would produce each year, noting that the waste would have to be trucked out through Long Island and the streets of New York City, where an accident or sabotage could cost 100,000 lives or more.

Then came the nuclear accident at Three Mile Island in 1979. Two months later 15,000 people demonstrated outside Shoreham and more than 600 were arrested. As a result of the accident, the federal Nuclear Regulatory Commission added requirements for nuclear operators to provide evacuation plans in concert with local and state government. But, of course, the only practical way to leave Long Island is via the bridges or tunnels into New York City, where the traffic bottlenecks are notorious in ordinary circumstances. Everyone living on the island knew that an emergency mass evacuation was simply impossible by land or sea. In February 1983, a month after Mario Cuomo took office as governor, he ordered state officials to reject any supposedly "safe" evacuation plan put forward by LILCO. Shortly thereafter the legislature of Suffolk County voted 15-1 that no safe evacuation plan was feasible.

Nonetheless, LILCO completed construction two years later, and by 1985 the reactor was prepared with radioactive materials and low-power tests were conducted. Presumably, LILCO officials assumed the plant and its burgeoning cost were simply too huge to be scuttled by any popular protest or government entity. They were wrong. That September, when Hurricane Gloria devastated Long Island, LILCO proved utterly incompetent in the recovery, to the surprise of few. Any remaining believers lost faith in the company.

The explosion at the Ukrainian Chernobyl nuclear plant in 1986 was another nail in Shoreham's coffin. Eight months later local grass-roots organizer Nora Bredes formed the Shoreham Opposition Coalition; by then 74 percent of Long Islanders opposed the plant, according to a *Newsday* poll. Dozens of groups, from the Sierra Club and the Audubon Society to the Farm Bureau, joined together, and the battle over Shoreham became, according to Dan Fagin's aptly melodramatic retrospective account in *Newsday*, "the defining political struggle of the Island's modern era," setting off a chain reaction "even beyond the nuclear industry whose decline it came to symbolize.... Like Frankenstein's monster, Shoreham eventually destroyed its own progenitor, the Long Island Lighting Co."

It was clear to Cuomo that, as an organization, LILCO had to be removed from the picture. In 1985, after extensive negotiations, the legislature passed the Long Island Power Authority Act, creating a new government entity, the Long Island Power Authority (LIPA), charged with acquiring LILCO's generating assets, including Shoreham. In 1989, Cuomo and the Chairman of LILCO agreed upon the complete decommissioning of the plant. That was when Richard Flynn became directly involved.

> Originally Shoreham came to me by (way of) someone who was very close to Governor Mario Cuomo. I had enormous admiration for Mario, both as an inspirational figure and as someone to work with. Anyway, this figure said to me, "What we want you to do is to buy Shoreham," and I said, "Look, you know, we're the New York Power Authority, not the New York Go-For-Broke Authority." But we negotiated an agreement with LILCO whereby we were paid like a contractor, fully reimbursed by LILCO for decommissioning it, and we successfully did that, achieving the governor's goal without destroying the Power Authority's economic self-sufficiency.... We solved the governor's problem in a cost effective way—that was a major achievement. LILCO fully reimbursed the Authority for all of its decommissioning costs."

The NYPA team dismantled the plant and its radioactive material was transported to a nuclear power plant in Pennsylvania. The job took until 1995 at a well-documented but reasonable price of $186 million, raising the total costs of Shoreham to a staggering $6 billion—roughly one hundred times the original estimate thirty years earlier.

But that was not the most outrageous outcome of the Shoreham debacle. In the end, Long Island ratepayers were made responsible for the plant's total cost, when in fact they had never needed it, if only because there were other solutions for the growing demand. Nonetheless, in accord with the agreement struck between the state and LILCO to pay for this fantastic fiasco, every Long Island customer had a 3 percent surcharge added to their monthly bill—*for thirty years*. Not a single watt of power ever came from the nuclear plant and yet, because of it, even today Long Islanders pay among the highest electricity rates in the United States.

The Revolutionary Ideals of Freeman

Flynn's successor was S. David Freeman, who was nationally renowned as a public power leader when Governor Mario Cuomo recruited him as CEO and President of NYPA in March 1994. (The titles of CEO and Chairman were divided during this period. Acting Chairman Thomas R. Frey, a former NYPA general counsel, was succeeded by Chairman Thomas G. Young in July 1994.) In the 1960s, at the Federal Power Commission, Freeman had ardently supported the expansion of public power; then, in 1967, President Lyndon Johnson asked him to formulate an energy policy for the nation. During his research for that task, he had an epiphany: "We were using energy the way the proverbial drunken sailor spends money. And it just hit me that part of the answer to the pollution problem as well as to energy planning was to move towards a more efficient use of electricity." That was exceptionally forward-thinking in the years before the OPEC oil embargo. But Freeman was also a pragmatist who did not allow the dream of a perfect solution to eclipse the reality of a good solution.

Freeman hailed from Tennessee, where President Jimmy Carter had appointed him to head the TVA, and where he deliberately focused more on energy efficiency than on growth; in fact, he cut off funding for eight nuclear plants that he considered too expensive. Later, he brought his "eco-pioneering" to the Sacramento Municipal Utility District (SMUD). One of his legacies there was a fleet of electric vehicles charged by solar panels. He started another program to give away 200,000 trees to the residents of Sacramento to provide shade for its many small houses in the brutal summer heat, quipping, "I think that I shall never see / An air conditioner lovely as a tree." And in Sacramento he launched a refrigerator rebate program aimed at energy efficiency, offering a $50 rebate for customers to replace their old fridges with new, efficient refrigerators—47,000 of them, as it turned out. "We buy the energy that our customers waste," he explained. SMUD was "buying energy from the cheapest source: our customers." SMUD's program was so successful that it was expanded exponentially in 1995. Perhaps the best symbol of Freeman's leadership in Sacramento was his installation of twenty acres of solar panels encircling a defunct nuclear power plant.

According to one environmental energy analyst, Freeman described himself as "a utility repair man who helps power companies make the transition from a dependence on fossil fuel and nuclear energy sources to a softer energy path comprised of conservation measures and renewable energy," because conservation, he observed, does not pollute, nor does it break down on a hot summer day. In New York, Freeman cut a somewhat unlikely figure, given his "studied sense of color," as one historian wrote, "with a Southern drawl and homespun expressions to match, often appearing on Manhattan streets in his trademark cowboy hat."

He wasted no time in emphasizing energy conservation. On his second day as CEO, he asked NYPA's trustees to cancel a pending $5 billion, 20-year contract to purchase 800 MW of capacity during the summer months from Hydro-Québec, which was preparing to build new dams in the far north of Canada that would impinge upon the native Inuit population who lived and hunted there. This followed the far more contentious cancellation of a controversial, year-around, 1,000-MW Hydro-Québec contract. The main reason Freeman cited was the sharp drop in demand and price in New York State in the years since the contract was negotiated. He attributed this drop partly to the economic recession and partly to the investments in energy efficiency introduced under Dyson and Flynn. The contract would have obliged the state to purchase three billion kWh that it did not need at a price that was above market, to then resell principally to Con Ed. "From a straight business point of view," said Freeman, "why would we want to buy more of this stuff when we've got enough?"

Under his watch, NYPA concluded ten-year contracts with the governments of Westchester County and New York City that froze rates and promoted energy efficiency; one of his programs focused on replacing coal-burning furnaces in New York schools, another on promoting electric vehicles. These were followed by other long-term government contracts. For his various efforts, Freeman was awarded the "Green Star" from the Environmental Action Coalition in recognition of the fact that he had doubled NYPA's financial commitment to energy conservation programs while also emphasizing electric transportation. Freeman also went through a complicated re-adjustment of rates for "replacement power," dating back to power contracts originally fulfilled by the old Schoellkopf plant. There is no question that Freeman accomplished a great deal in his short year and a half with the Authority.

Economic Transformations

In 1994, when George E. Pataki became New York's first Republican Governor in twenty years, he chose Clarence D. "Rapp" Rappleyea to lead NYPA as chairman and CEO, while Robert G. Schoenberger became its president and chief operating officer. Rappleyea had been in the state Assembly for twenty-two years and served as its minority leader for twelve (a period that included Pataki's eight years as an assemblyman). In July 1995, NYPA's trustees elected Rappleyea, who the *New York Times* described at the time as "a political mentor to Gov. George E. Pataki." Thanks to his "personable manner and uncanny knack for remembering names," wrote one observer, "Rapp, as he was universally known, established an instant rapport with Power Authority employees. The fact that he was highly regarded on both sides of the aisle in Albany would also prove invaluable during his eventful 5½-year tenure, as would his zest for competition."

At the urging of the Legislature, Rappleyea dramatically expanded the Authority's economic development role by implementing a new "Power for Jobs" program that had evolved from Dyson's "Juice for Jobs" and Flynn's more robust "Power for Jobs." Now, the state's Economic Development Power Allocation Board would apportion low-cost power (400 MW—half from the FitzPatrick plant, half from competitive bidding for energy from other companies) on the basis of much more specific criteria. The new, three-year "Power for Jobs" program was such an immediate success that the power allocation was raised to 450 MW and repeatedly extended for years, ultimately serving some 600 businesses and nonprofit entities. By early 2006, "Power for Jobs" was credited with securing more than 300,000 jobs, and continued adding to that number for the next five years.

It was also on Rappleyea's watch that the entire structure and nature of the state's monitoring and pricing systems for electricity were transformed. Over the previous three decades under the operational control of the Power Pool, as demand grew for more generation, the price of electricity in New York had climbed, first because of the OPEC oil embargo, then from construction cost overruns of numerous large new nuclear plants, and then from the cost of decommissioning Shoreham. Adam Smith's "invisible hand"—the market force of competition that can reduce prices and improve efficiency—was severely constrained by the Power Pool's arcane system of calculating supply and demand, which quickly became obsolete in the age of the Internet. As a result, by the mid-1990s New York's retail electricity rates were among the highest in the country, according to cost-benefit analysis. Several industrial customers lobbied for relief from the high prices, while others economized by installing their own generators and leaving their suppliers.

Much the same problem was impacting the whole country. Congress had tried to remedy the situation nationally with the Energy Policy Act of 1992, which "greased the skids for greater competition within the electric supply industry," reported one industry historian, with the aim of attracting competitive new utilities to generate power and thus lower prices. The Act also stipulated that transmission lines must be made available to competing power plants, and that all the generating companies must provide real-time information about their power production and transmission. But despite the new law, the entrenched, vertically-integrated utilities dragged their feet on opening their markets, continuing to stifle competition; after three years hardly any new generation had been launched.

It naturally fell to the Federal Energy Regulatory Commission (FERC) to implement the Energy Policy Act and achieve the industry restructuring that the law called for. In April 1996, FERC issued Order 888, which mandated unbundling the utilities' generation functions and provided open access to transmission facilities. Order 889, issued in 1996, instituted the Open Access Same-Time Information System (OASIS), an Internet-based system with which the power companies can see where and when transmission facilities are available and can reserve blocks of time on those facilities. More than anything else, it was this technology that made the evolution from power pools to real control centers possible.

The two orders also called for the development of Independent System Operators (ISOs) and Regional Transmission Organizations (RTOs) to act as autonomous grid operators who were *not* beholden to the investor-owned utilities and could provide the flexibility necessary to add new sources of generation as they were developed under FERC's new Open Access policy. Without a monitoring and auctioning system that was truly independent of the utilities, it had proven almost impossible to attract utility competitors, especially in regions where a single company controlled both generation and transmission.

The New York ISO (NYISO), like the others created since 1996 (mid-Atlantic region, California, New England, the Midwest, and Texas), has become an independent, nonprofit control center balancing supply and demand and maintaining voltage, while monitoring for contingencies and managing reserves. At the same time, NYISO provides a number of markets in which generating utilities and their customers can do business. The three most fundamental are the Day-Ahead Market, the Real-Time Market, and Bilateral Transactions. In the Day-Ahead Market, the ISO in effect sets the price by choosing which bids they accept twenty-four hours in advance. This allows the power companies to anticipate how many generators they will need to operate to meet the next day's demand, since starting up a generator can take hours. The Real-Time Market, which represents only 2 percent of all the transactions, serves to balance and fine-tune the supply, depending upon system conditions and unexpected demand. Bilateral Transactions, which represent half the sales that the NYISO supports, are negotiated outside of the marketplace but are entered into the system to make transmission service available.

This bid-based system, explains NYISO, employs "a uniform clearing price auction to provide a common price at each location for all buyers and sellers. This type of auction creates an incentive for producers to lower their costs and bid competitively. They also provide transparency to the marketplace, as all participants are aware of the value of energy." Their system uses "sophisticated software that simultaneously determines how to serve the load, utilizing generators with the combined lowest total production cost." What seems most remarkable is that these computer-driven NYISO auctions take place continuously, every five minutes, twenty-four hours a day.

Today more than 400 companies make bids through NYISO, which serves wholesale customers that purchase a minimum of 1 MW (enough to power 600 to 1,000 homes). Their price includes all the costs of maintaining the system, including reimbursement to the Power Authority for use of its transmission lines, operators, and service, along with the NYISO's operating costs and other flat-rate charges for maintaining the system. The Power Authority also buys and sells electricity with the other utilities. NYISO provides several ancillary markets as well. Finally, the financial markets play a role in improving market efficiency in New York State, allowing purchasers to lock-in a congestion cost payment, which helps to fulfill the FERC requirement of providing "firm" transmission service. And in the Virtual Market, investors arbitrage the day-ahead and real-time price differentials, which has also been shown to increase market efficiency.

Under the NYISO, consumers can buy electricity from suppliers anywhere in the state or beyond instead of being limited to local or regional suppliers. The intention from the start was to create competition that would drive prices lower because, as Rappleyea said in 1997, "Competition works best when there is plenty of it. You don't increase competition by eliminating competitors—and that includes public power." Eugene W.

Zeltmann, NYPA's president and chief operating officer at the time (and later president and CEO) declared that "it's our responsibility to seek collaboration—not confrontation—with the state's investor-owned utilities and others involved in the reshaping of our industry."

It was on December 1, 1999, that control of the flow of power across New York passed from the Power Pool to the NYISO. Rappleyea, speaking at the ceremonial transfer that day in his capacity as chairman of the Pool's Executive Committee, described "the passing of the torch to the New York Independent System Operator" as "a major milestone on the road to a competitive marketplace for electricity in the Empire State."

Selling NYPA's Nuclear Plants and Moving Forward

From the time he took charge of the Power Authority, Rappleyea, like his predecessors, had to devote considerable attention to NYPA's nuclear plants, Indian Point 3 and FitzPatrick, where there were technological problems that required a higher level of expertise and leadership than their combined workforce possessed. William J. Cahill, Jr., a forty-year veteran of nuclear power and NYPA's chief nuclear officer, improved the plants' operations and finances. But in 1996, after negotiations fell through with Entergy Corp. (a large New Orleans-based utility) to operate the two plants, Rappleyea redoubled the efforts to improve the plants' performance, and by 1998 they were setting combined production records. That again attracted the interest of Entergy, this time with an offer to purchase the plants. After several months of exclusive negotiations between NYPA and Entergy, another potential buyer, Dominion Resources of Richmond, Virginia, made a greater offer for the two plants. For the next month, the two bidders raised the value of the sale considerably, to the benefit of New York, and in the end, it fell to the trustees to consider all aspects of the two offers and choose that which best served the interests of the state's residents.

The trustees, including Rappleyea, chose the bid from Entergy, which included $636 million for the two plants, another $171 million for nuclear fuel on hand or on order, and other elements. The total price came to $967 million; per kilowatt of generating capacity, the price of $536 was fourfold that of the previous record for a nuclear plant. And most of the 1,500 NYPA employees trained in nuclear work at the two plants transferred to Entergy with no loss of pay or benefits. The sale was completed in March 2000 and with that, the Power Authority was no longer in the business of nuclear power.

In August 2000, the last full year of his tenure, Rappleyea announced the Power Authority's next project, dubbed "PowerNow!," to build five small, clean power plants along the rivers in New York City, and to do it in record time, in response to dire predictions of power shortages and brownouts as early as the following summer. Two more small plants, on Staten Island and in Brentwood, Long Island, were added soon after. These plants, fired by natural gas, were equipped with advanced technology—$85 million worth—to reduce both noise and pollution. To address the small quantity of pollutants that would be emitted by the New York City plants, NYPA introduced new technologies at other sources of pollution in New York City that reduced the waste gases by a commensurate amount, calling this a "zero net emissions program." In less than a year, the seven plants in the Bronx, Brooklyn, Queens, Staten Island, and at Brentwood on Long Island were adding a total of about 460 MW to the grid, in close proximity to the demand, and sure enough, the added power made a critical difference during the first two hot summers of operation and at various times in the years to come. During a 2001 heatwave, the *New York Post* observed that "[Governor] Pataki and NYPA scurried to get the turbines up and running. Thankfully, they succeeded."

The day before Rappleyea left office in January 2001, NYPA's trustees surprised him with an unexpected honor: they named the Power Authority's main office in White Plains the Clarence D. Rappleyea Building. As Governor Pataki noted of "Rapp" that day, "his strong leadership guided the New York Power Authority through a period of sweeping changes in the electric utility industry."

CHAPTER ELEVEN

A MODEL FOR BEST PRACTICES

Reducing Clients' Energy Use

In order to make more electricity available, one can either generate more or use less by improving energy efficiency. In the last two decades the New York Power Authority has devoted almost as much effort to the latter as to the former. Under the direction of the state government and in cooperation with a few key agencies and organizations, the Power Authority has implemented a wide array of projects to improve energy efficiency, many related to lighting, heating, and cooling. Similar efforts are made by municipal and state utilities across the country. The same can rarely be said of private utilities, even those that have expanded their conservation efforts. They may offer customers "useful tips" to reduce usage, but the fact remains that it is generally in their shareholders' best interests if their customers are *not* energy efficient. The Power Authority goes well beyond useful tips, thanks partly to its key partner in efficiency, the New York State Energy Research and Development Authority (NYSERDA).

NYSERDA is a public benefit corporation created in 1975 by Governor Hugh Carey after the first Oil Embargo; since 1998 it has been funded by a miniscule charge on the bill of every customer of an investor-owned utility. NYSERDA collaborates with businesses, industry, the federal government, academia, the environmental community, public interest groups, and energy market participants to reduce energy consumption and greenhouse gas emissions. It also provides information and analysis, programs, technical expertise, and funding aimed at developing renewable energy while protecting the environment and creating clean-energy jobs. In many cases, the Power Authority seeks solutions from NYSERDA's researchers, who at other times propose projects that NYPA adopts.

NYPA's efforts to reduce its clients' usage had begun in 1980 with Dyson's Button Up program (followed by the similar Watt Busters program), which offered customers of the state's municipal systems and rural cooperatives free home energy audits, and which had a response rate far above similar programs that the state's investor-owned electric and gas utilities achieved. Then, in 1990, the High Efficiency Lighting Program, or HELP, program was announced, targeting the Power Authority's governmental customers in New York City and Westchester County. The Authority took charge of each project, determining and designing the lighting required at customers' facilities and then providing contractors to install energy-efficient lighting and to affirm quality assurance. What was perhaps most remarkable was that the Authority financed each job, recovering all its costs from the savings on energy bills, whereupon the customer kept the savings. Over the next three years, HELP was expanded to state, municipal, and county facilities as well as public schools and community colleges statewide.

While these retrofitted lighting projects provided significant savings, there were bigger opportunities waiting in the field of heating, ventilation, and air conditioning (HVAC). All the HVAC technologies had made tremendous strides in efficiency, and upgrades performed by NYPA benefited government buildings and public schools. If co-fired boilers were installed, they could provide hot water along with heating. As well, the state began issuing New York bonds to pay for the conversion of generators from coal to oil and from oil to gas.

By the 1990s, NYPA could offer its customers "soup-to-nuts" efficiency services: evaluating current usage and existing equipment; designing and testing solutions and replacements; negotiating prices for new equipment; and actually executing the engineering and installation, although additional outside contractors were usually involved as well. But the most unique feature of these "all-in" efficiency services was then and still remains NYPA's ability to finance them and then recover the costs from the savings. Having the fiscal latitude to carry out such creative financing allows the Power Authority to seek out and find the paradigms in efficiency that can then serve as "best practice" models for all the power companies, including the investor-owned utilities; or at the very least to expose the deficiencies of those that disregard them.

The NYCHA Energy Efficient Refrigerator Program

Perhaps the most ambitious and challenging energy efficiency initiative the Power Authority ever undertook was for the New York City Housing Authority. NYCHA, as it is known, was created in 1935 as a slum-clearance project financed by bonds. Its First Houses (as they were named) were developed on Avenue A on the Lower East Side. NYCHA's most dramatic growth came in the postwar era under the autocratic reign of none other than Robert Moses, who replaced the infested, six-story wooden tenements of the 1900s with brick high-rises twice as tall. NYCHA's stated mission is "to increase opportunities for low- and moderate-income New Yorkers by providing safe, affordable housing and facilitating access to social and community services."

Today, the nation's largest landlord is too vast and scattered to describe clearly without statistics, which always threaten to treat families as mere numbers. In 2014, NYCHA housed almost 5 percent of New York City's population, with some 400,000 residents living in more than 180,000 apartments. These are dispersed among 2,563 buildings with 3,330 elevators. The Power Authority provides NYCHA with low-cost electricity for all of its 334 separate developments, long known as "the Projects."

One can only imagine how much electricity is wasted in all the city's thousands of residential buildings, NYCHA-owned and others; how many pointless trips the old elevators make, how many leaking faucets keep the building's pumps working, how many lights and televisions are left on, and how many inefficient refrigerators, some decades old, are still operating. Such wastefulness is usually paid for by the individual tenants, but for NYCHA apartments the electricity bill is paid by the state, which therefore has both the motive and the means to address that waste. But by what means, exactly?

Almost as soon as S. David Freeman, became NYPA's CEO in 1994, he proposed a much more robust version of the refrigerator rebate program he had introduced in Sacramento. After lengthy research and calculation, in 1996 NYPA and NYCHA launched a program of Herculean proportions aimed at replacing the refrigerators in 184,000 NYCHA apartments with more efficient units. That was roughly enough new refrigerators to cover thirty-six football fields, leaving a similar mass of old refrigerators to be disposed of as part of the job. NYPA organized the procurement of the most economical refrigerators directly from the manufacturers (at a substantially reduced wholesale price) and entirely free of charge to the residents, with most of the physical effort shouldered by Power Authority employees. As the program progressed and manufacturers made improvements in efficiency, NYPA continued to ensure that NYCHA received the most efficient models available, which used one-third to one-half as much electricity as the old units. Reducing the waste from all those aging, inefficient fridges would be as good as generating additional power. But could it possibly be less expensive? This vast operation required a $72 million investment and took a large team a very long time; as planned, the program took nine years.

Eric Alemany was a key player in the execution of the program. He recalled that "it was the biggest job I had ever undertaken by far." He had joined NYPA's Energy Efficiency group in 1998 as an Assistant

Conservation Engineer and spent six years carrying out the job, replacing refrigerators every day in the low- to moderate-income households, including those of some of the state's lowest-income residents. In each case his team brought a full-size refrigerator to the apartment at a prearranged time, removed the (often very) old unit after the chilled contents were hastily removed, delicately maneuvered it into the hall, then installed the new refrigerator, and left with the old one and the packing materials. It's hard to imagine, but on average they replaced ninety refrigerators every day; harder still to imagine are the thousands of domestic microworlds that they experienced in those apartments.

When the project was completed in April 2005, it had succeeded in reducing the Housing Authority's electrical consumption by 110,790 MWh (megawatt hours) per year, freeing 13.8 MW. That is enough to supply at least 13,000 additional households. The "NYPA/NYCHA Energy Efficient Refrigerator Program" paid off its own initial costs years ago and continues to save the state and the Housing Authority about $7.7 million every year. In the decade since the project ended, the Housing Authority has continued with its own replacement plan, since the newest refrigerator technology (linear-motion compressors, vacuum panel, and black carbon insulation) delivers enough energy savings to warrant beginning a new round of exchanges.

From the start of its planning, the Power Authority had a greater responsibility than just cost-cutting and energy-saving. Because after all the new refrigerators were installed, exactly what was to be done with the old units? Clearly NYPA couldn't leave 146,000 moldy, rusted fridges on the streets for the Sanitation Department to dispose of. The state-owned power company had to obey the law, at the very least. In fact, as an instrumentality of the state, NYPA has a greater obligation: to address the massive disposal problem *in an exemplary fashion* by determining exactly what the best practice was in this situation, a role that has no parallel in the profit-driven market.

This "higher calling," if you will, is possibly the greatest distinguishing factor of NYPA and other diligent publicly owned utilities. This principle has hardly been achieved to perfection in all places, but for the last forty years a hot spotlight has usually caught up with government agencies that violate environmental policy or fall short of "best practices." Still, how are the best practices for any given project determined? Although experience offers guidance, there really are no shortcuts. At NYPA, after thorough assessment by teams of specialized engineers, in-house or contracted, deliberate, trial-and-error approaches are assessed and demonstrated in consultation with NYSERDA and other organizations.

In the case of the 146,000 discarded refrigerators (some 38,000 newer units were either cannibalized for parts or swapped for older ones), NYPA planned the recycling operation from the outset, with environmental guidelines the leading consideration. For example, more than 18,000 lbs. of ozone-depleting R-12 refrigerant were reclaimed, sparing the ozone layer some significant degradation. The scale of the whole operation is difficult to grasp. The cardboard boxes alone from all the new refrigerators weighed 2,000,000 pounds. In addition, over 24 million pounds of steel, 160 tons of aluminum, and almost 30 tons of copper were recycled. Perhaps most significantly, the refrigerator replacement program reduced the emission of greenhouse gases by a whopping 77,509 tons annually, the equivalent of removing almost 15,000 vehicles from the road.

NYPA's Energy Efficiency Division (formerly called Energy Services) has continued replacing refrigerators at Municipal Housing Authorities around the state, from Buffalo to Lake Placid to Tupper Lake. But this is only part of the overall energy efficiency effort. In all, at high schools, libraries, waste treatment plants, airports, hospitals, and public facilities statewide, NYPA has spent a total of $1.5 billion at more than 4,000 sites, which produced annual savings of $148 million and reduced demand by 1,183,000 MWh. As a result of NYPA's energy efficiency projects, the energy equivalent of 1.6 million barrels of oil was not used annually by the state to supplement its hydropower, eliminating 260,000 tons of greenhouse gas emissions annually. In 2013 alone, the Power Authority spent $281 million on efficiency programs in 336 facilities, resulting in savings of $13 million and burning almost 170,000 fewer barrels of oil.

BuildSmart NY

At the end of 2012, Governor Cuomo issued Executive Order 88, mandating a 20 percent improvement in the energy efficiency performance of state government buildings by April 2020. Administered by the Power Authority, EO-88 is the centerpiece of BuildSmart NY, a state program led by NYPA's President and CEO, Gil Quiniones. This is more ambitious than anything undertaken before, given that the state owns or controls some 212 million square feet of real estate, including universities, prisons, mental health hospitals, and office buildings, among others. The totality of these buildings consumes a staggering 60 percent of the total energy used within New York State, emitting *half of all the state's greenhouse gases*. So reducing those totals by 20 percent in the next four years will definitely have a dramatic impact. It is a bold initiative, built confidently upon the historical success, environmental *and* economic, of previous energy efficiency projects. More than 90 percent of the state's energy bill goes to the state and city universities (SUNY and CUNY), the Department of Corrections, the Offices of General Services and of Mental Health, and the Metropolitan Transportation Authority. Naturally, then, the emphasis of EO-88 is upon improving efficiency at these agencies.

The metric by which success is assessed is called Source Energy Use Intensity, which measures all energy use per square foot, taking into account what it costs to provide the specific fuels consumed to power each building: coal, oil, natural gas, solar, wind, or hydro. That is because nationally, for every BTU of useful electricity at the point of end use, 3.34 BTU of energy are consumed to generate and deliver that electricity.

The process begins by establishing baseline and benchmark levels with energy audits of every government and agency building involved, to measure subsequent progress. In many cases this requires "submetering"—adding electric meters to different parts of a building to determine where and how most of the power is used. Then NYPA engineers look at how and why a building's systems are operated and maintained as they are, and identify ways to improve overall building performance, called "retrocommissioning." Energy master plans are designed for the larger campuses, and a clear-cut schedule for progress on each project is established. At the same time, complex financial planning began, to fund each major stage.

In 2013, the BuildSmart NY Benchmarking Report addressed conditions at the more than 16,000 state-owned buildings, many of which required submetering to identify the baseline usage by individual buildings instead of entire "master-metered" campuses. "Few CUNY and SUNY college campuses are metered at the building level," the report emphasizes, while also noting that the best practices prescribed by the Power Authority's team should extend beyond government buildings. "From an economic perspective, BuildSmart NY is also intended to catalyze the New York State marketplace, by serving as an example for private building owners, demonstrating new technology and applications, and providing opportunities for the energy efficiency industry to provide innovative solutions to State agencies."

The Poletti Project

Some of the best practices that the Power Authority has demonstrated go beyond projects that are strictly electrical. In fact, the ultimate destiny of the Poletti Power Project in Astoria, Queens, which had begun generating power in 1977, speaks volumes about NYPA's efforts to practice sustainability in all its undertakings and to innovate solutions for specific, unprecedented problems.

By the late 1990s, the dual-fuel, 885-MW plant fell far short of the cleaner technology available. The region downwind from the smokestacks of several companies was known as "asthma alley" because of its statistically high rates of lung disease, especially among children. The Poletti Project was wrongly blamed for most of this when, in fact, it was cleaner than some of the older investor-owned plants nearby, and it was much cleaner during the summer months when it burned natural gas. Nonetheless, in winter, when gas supplies were reserved for residential users,

Poletti burned #6 oil, producing predictable environmental harm, according to the Natural Resources Defense Council. Local environmental groups began protesting and petitioning, led by CHOKE (Coalition Helping Organize a Kleaner Environment) and the United Community Civic Association.

The New York Power Authority, under Rappleyea, had already announced its intention to build a 500-megawatt beside the Poletti Plant on the 47-acre site owned by Con Edison, joining the NRG and US Powergen Plants in providing most of New York City's electricity. NYPA's new facility would be a combined-cycle plant, with two turbines driven by the combustion of natural gas (with low-sulfur oil as a backup) and one driven by steam boiled with the heat recovered from the first two, making the whole plant 50 percent more efficient than conventional units. The steam would then be cooled and recycled, obviating the need for water from the East River.

With its combined-cycle technology, the new plant would be the most efficient in New York City's history. Its sophisticated controls would enable it to meet federal emissions standards at least equal to any then in place for the nation's power plants. Its air-cooled condenser would eliminate concerns about aquatic life and water quality. With its location beside Poletti, the new plant could use the existing infrastructure, including a natural gas supply, fuel storage facilities, and the complex transmission facilities.

In 2002 Governor George Pataki announced an agreement under which the Power Authority would close the existing Poletti Project no later than 2010. In return, opponents agreed to support the Authority's application to build the new plant, which was quickly approved. The Authority pledged to: impose operating restrictions at the existing facility until that plant was eventually closed; increase its sizable investment in energy efficiency projects in New York City; and target $2 million for projects to clean the air in Queens. All these initiatives were completed or in progress by early 2006. Still, local officials and environmental leaders had argued that, given all the Queens generators, another plant, no matter how clean or efficient, would only add to their burden, and this local hostility fed in to the mistrust during negotiations about the demolition process for the Poletti Plant.

Construction began on the $650 million 500 MW plant in 2002, where commercial operation began at the end of 2005. It had been thirty years since the Authority's purchase from Con Edison of Indian Point 3, which had been the first source of power for the Authority's governmental customers in New York City and Westchester County. The new 500-MW plant's importance had been underscored earlier in 2005 when the Authority reached innovative power-supply agreements with New York City customers that extended through 2017. The agreements created a collaborative process that called for the customers to participate extensively in the Authority's planning and decision-making and to choose from various payment options. As the 500-MW plant began operation, the Independent System Operator was forecasting the need for still more resources to serve the downstate area within the next few years. Fortunately, the new privately owned Astoria plant came online soon after, with NYPA agreeing to buy its output.

In 2010, as planned, NYPA's Poletti Project was formally decommissioned. But that was not enough for local residents, who had seen plants shut down in the past, only to be started up again years later. So one enormous chore remained: the entire plant had to be removed. By this time there had already been years of discussion about how the process would unfold. To the casual observer, who might imagine a wrecking ball slamming into the side of the enormous structure, one can instead suggest picturing an accelerated film of its construction being played backwards. Under the meticulous scrutiny of local groups—especially the United Community Civic Association—concern had already been vociferously expressed about possible asbestos contamination and other pollutants, the traffic disruptions caused by thousands of truckloads hauling all that material away, and more. Older residents vividly recalled what they viewed as the heedless and relentless parade of trucks that delivered all the material for the site in the early 1970s when Con Ed began construction. So once again NYPA had been challenged to demonstrate the best possible practices, and spent years developing its plan, along with $30 million, to carry it out.

Ed Birdie is the Power Authority's Director of Southeast New York Community Affairs in NYPA's Public, Government, and Regulatory Affairs Department. He recently explained that the five-year process that began when generation ceased in 2010 with permanent lock-out from the facility was followed the next year by the salvage and sale of some of its equipment at market value, and then detailed design of the deconstruction plan that would ensue. NYPA engineers went back to the original construction plans followed by Con Ed when it began building the facility in the early 1970s, before NYPA took over and completed the job. The cranes brought in to remove huge steel beams were identical to the cranes that had put them in place. This "reverse construction" assured a controlled and methodical process that would produce a minimum of pollution. As well, every scrap of disassembled metal, including the huge oil tanks, was sold in bulk as recycled material.

As for transporting that bulk, community organizers pressed NYPA to use barges along the East River rather than trucks along the streets, but after months of consideration, that option was rejected because of its cost. Instead, the Power Authority sought and received the endorsement of New York's Department of Transportation and local groups for a route through designated streets only, during specified hours, by trucks equipped with exhaust suppressors. By the end of 2014, the last pieces of the giant plant had been removed.

Following the methods of the US Climate Registry, in 2009 the Power Authority recorded the first comprehensive baseline analysis of its own carbon footprint, in which CO_2 emissions from NYPA's facilities (excluding vehicles) totaled 2,821,965 metric tons. By 2011, just by closing the oil- and gas-burning Poletti plant in 2010, NYPA had reduced those emissions by 28%.

Demonstrating Best Practices Internally

The responsibility of a government agency or instrumentality to perform work in the most environmentally sound manner may seem obvious today, but it has only been articulated in the last thirty years or so, and acted upon with varying degrees of success and abject failure by different agencies. In the case of the Power Authority, of course, given its strong public stand on energy efficiency and sustainability, any lapse in its own practices is perceived as corporate hypocrisy. While offering to improve the efficiency of its business and governmental customers' buildings, NYPA has an obvious obligation to demonstrate best practices at the seventeen-story Rappleyea Building in White Plains, which had been built in 1980. Almost half of NYPA's 1,600 employees share its (total) 420,000 square feet of office space, together with a number of tenants. In 2002, the Authority completed large-scale efficiency and environmental improvements that included a much-improved HVAC system, ultra-high-efficiency lighting, and insulation that thoroughly isolates the building atmospherically. As much as 15 percent of the building's power comes from a 200-kilowatt natural gas fuel cell, which generates electricity from chemical reaction rather than combustion. These efforts cost some $3.4 million and reduced the building's overall energy use by 50 percent.

The building soon met the guidelines prescribed by the US Green Building Council, which is made up of some 12,000 member organizations and 193,000 Leadership in Energy and Environmental Design (LEED) professionals. It provides certification and has guided the design, construction, operations, and maintenance of more than 54,000 projects globally. As a result of NYPA's retrofitting on the high rise, the Green Building Council presented its LEED Gold-EB (Existing Building) Award to the White Plains Office in 2007, and in 2013, it was the first building in the state to be recertified. A vast new warehouse in Niagara was also certified as meeting LEED Gold standards. Nearby, at the Niagara Power Vista, a photovoltaic or "solar" array was added on the rooftop, and at the Blenheim-Gilboa Visitors Center a wind turbine and solar array were installed on the grounds as both a source of power and an educational exhibit.

But the larger goal of perfecting and maintaining sustainable practices is a never-ending process that progresses from obvious problems to smaller issues. So, on the grounds surrounding the Rappleyea Building, the landscaping is now done with local plants that need less water; the cleaning products used by the janitorial staff have been replaced with environmentally friendly alternatives; and NYPA has developed green guidelines for renovations that provide more natural light while improving indoor air quality and workplace efficiency. These actions, along with all the others, have saved at least two million kilowatt-hours per year.

As scrutiny of workers' habits became more refined, the head of NYPA's sustainability programs, Kerry-Jane King, aimed at reducing the total amount of trash produced every day in the building—notably, plastic bottles. With that, water coolers were installed on every floor; employees were issued metal thermoses for their beverages and encouraged, if only by peer pressure, to use them instead of plastic bottles. As well, NYPA organized employee Green Teams and annual Green Expos at its largest sites, urging workers to "Get Involved!" with activities ranging from green-themed eco-cinema to expanded recycling programs. And new wastebaskets appeared throughout NYPA's offices to improve the sorting of recyclables.

Socially responsible sustainability does not end within the corporate workplace. In 2011 NYPA introduced flexible schedules for employees, along with preferential parking for hybrid and electric vehicle owners and carpoolers, to reduce highway congestion and thus greenhouse gases. Almost immediately, the employees' commuting patterns began changing: free now to avoid traveling by car or train during peak hours, hundreds shifted their schedules, reducing daily peak congestion on the roads and rails.

Eco-Transport

Transportation presents the Power Authority with another opportunity to research and implement best practices, in collaboration with NYSERDA, by participating in national, state, and regional programs to promote the development and demonstration of electric, hybrid, and plug-in hybrid-electric vehicles. Since 2003, NYPA has approved loans for municipal and cooperative systems to purchase such vehicles for their fleets; to date, forty-six vehicles have been provided to twenty-one municipalities. As well, NYPA has supported the development of hybrid-electric buses. After New York City Transit tested them, they ordered and received 815 buses, and then ordered another 850. This program won the EPA's Clean Air Excellence Award.

In their early years, electric vehicles (EVs) were barely practical for NYPA because of their limited range; they often could not travel the full distance from one NYPA facility to another. Since then additional charging stations and improved vehicle performance have made them much more useful. NYPA also started providing nitrogen gas for its vehicles' tires, further improving their mileage. Now, each year NYPA replaces more of its own fleet with alternative-fuel vehicles and multiplies the number of electric charging stations for cars, trucks, buses, vans, and off-road equipment. In 2011, the Power Authority co-funded the purchase of fifty Chevrolet Volt extended-range electric vehicles and ten Ford Hybrid Transit Connect electric vans for nine New York City departments. NYPA also contributed to the purchase of hybrid shuttle buses for the villages of Tarrytown and Hastings in Westchester County to offer outreach services to senior citizens, all the while evaluating the vehicles' performance. In all, NYPA has helped provide more than 1,250 electric and hybrid electric vehicles, which have traveled more than 11 million miles.

One of the greatest obstacles to the widespread use of electric vehicles, of course, is the scarcity of charging stations. This is especially apparent in the greater metropolitan areas where EVs are best suited, given the limited range of their batteries. Charging stations require space dedicated to their exclusive use and a proximate source of high-wattage electric power, which may require some underground installation. Even a "quick charge" of a Prius at 240V takes 90 minutes. Traditional gas stations have not been quick to add charging facilities. In fact, it is exactly the kind of problem (like Wi-Fi in public places) that *laissez-faire* capitalism is unlikely to solve by itself.

In 2012, Governor Cuomo awarded $4.4 million to ten companies, municipalities, and other entities to install more than 325 new electric-vehicle charging stations across New York State. This was only a prelude to his "Charge NY" initiative announced the following year, which aimed to expand dramatically the prospects for EVs in the state, first by examining and eliminating any regulatory obstacles, then by awarding competitive contracts to New York companies to site new charging stations in parking lots and garages, including transportation hubs, mostly in the New York City area. The construction is financed with $50 million from NYPA, NYSERDA, and tax credits, and the target is to create 3,000 charging stations across the state and put as many as 40,000 plug-in vehicles on the road by 2018. Some of these will incorporate a design approved by the Electric Power Research Institute for solar-assisted vehicle-charging systems.

Renewable Energy Across the Nation

Of course, the New York Power Authority is only one public utility that is earnest in its research and application of best practices. Across the country more than 2,000 community-owned electric utilities now serve more than forty-seven million Americans. Operated by local governments, they provide reliable, responsive, not-for-profit electric service, just as the first one did in 1880, when Wabash, Indiana, (pop. 3,800) proudly lit its courthouse square with four arc lamps. Even today, most public power communities are small, with their utilities serving 3,000 or fewer customers. and they are directly accountable to the people they serve through local elected or appointed officials. Los Angeles, San Antonio, Austin, Seattle, Orlando, and other large cities across the US also operate publicly owned electric utilities.

What they all have in common, in contrast with every investor-owned utility, is a strong community-based incentive to provide the cleanest possible energy at the lowest possible price. Energy efficiency is one of the indispensable components to achieve that goal. It is only thanks to energy efficiency that since 2010, electricity use by US consumers actually dropped to its lowest level in more than a decade, according to the Energy Information Administration. There is no doubt that the recession played a role in this, but it is nonetheless quite an achievement in an era of such growth in new machines and electronics, and it is the result of more efficient homes, appliances, and technical devices. The best energy efficiency practices developed by all the utilities are shared through an industry-wide consortium, the Electric Power Research Institute (EPRI), for which NYPA's President and CEO Gil Quiniones became Chairman in 2015.

But, of course, another indispensable component is the accelerating movement toward renewable energy in New York, the United States, and the world. Thanks to its primary production of hydropower, 70 percent of the Power Authority's energy output is from a renewable source. Efforts to raise that percentage through additional solar, wind, and biomass generation have accelerated dramatically since 2012. NYPA has installed solar panels on SUNY campuses and hospital rooftops across the state, installations that repay NYPA from the savings that are realized. Until the Northeast as a whole adapts to the more flexible Smart Grid technology that can adjust with the variable output of solar and wind generators, NYPA is unlikely to take the lead in developing any large-scale renewable energy projects in the near future. For the time being, more of its efforts are devoted to statewide projects in energy efficiency.

Investor-owned utilities are being pressed by New York and some thirty other states to develop more renewable energy, on a schedule, with a Renewable Portfolio Standard (RPS). In 2004, the New York Public Service Commission (PSC)—the Commission created by Governor Charles Evans Hughes almost exactly a century earlier—instituted such a standard, and later set a target of 50 percent renewable energy generation in the state by 2030. This was not applicable to NYPA or the state's municipal utilities, though most have adopted similar programs. The investor-owned utilities can reach the target by adopting large-scale generation from biomass, fuel cells, photovoltaics, or ocean, tidal, and wind power. They can also install customer-sited systems such as solar hot water, fuel cells, methane digesters, and perhaps the most promising, wind turbines. According to the New York Independent System Operator (NYISO), between 2005 and 2013 alone, the generating capacity of wind-powered projects interconnected with the New York grid grew from 48 MW to 1,730 MW, with another 2,000 MW of wind power in development. In some cases, utilities may either own the renewable energy facilities or buy electricity from them.

Nationally, there are other significant reasons for hope. In 2012, 12 percent of the electricity produced by US utilities came from conventional hydropower and renewable energy resources, up from 7 percent in 2001. With the boom in shale gas production, prices for natural gas began falling in 2006. Natural gas has since become the choice of more US utilities for new power plants, and even existing plants that convert to it, in anticipation of recent EPA regulations; it is also well-suited to back up or fill in the gaps created by variable wind and solar production. Natural gas emits roughly half the amount of carbon dioxide as coal, far less sulfur dioxide, and none of the particulate discharge so harmful to human health. Electricity generated from natural gas has almost doubled since 2006 in the United States, and just since 2010 more than a hundred coal-fired plants across the country have shut down. Nonetheless, about 37 percent of US electricity is still generated from coal.

Another external inducement to best practices is the Regional Greenhouse Gas Initiative (RGGI), a cooperative effort between nine Northeastern and mid-Atlantic states to employ a market-based regulatory program to reduce carbon emissions. As of 2014, RGGI set a regional "budget" cap of ninety-one million short tons of CO_2 from power plants, declining by 2.5 percent annually thereafter. The states are allowed to sell their emission allowances to invest in energy efficiency, renewable energy, or other consumer benefit programs. New York, an active member of RGGI, sells its allowances, and since 2012 has used $100 million of the proceeds each year for Governor Cuomo's program called Cleaner, Greener Communities, a statewide initiative to reduce greenhouse gas pollution through smart growth planning and sustainability. RGGI proceeds also support sixteen large-scale, cost-shared projects increasing energy efficiency, renewable energy, and carbon abatement. These projects address regional priorities, such as converting oil-fueled heating boilers to natural gas at an International Paper plant in Ticonderoga, or construction of a model net-zero energy building at the SUNY College of Nanoscale Science and Engineering in Albany.

All these efforts by the state, the regions, and the public utilities serve to combat climate change, a cause to which President George H.W. Bush made a firm commitment in 1992—most recently reaffirmed at the Paris Climate Conference (COP21) in December 2015—while undertaking little or no action nationally on greenhouse gases. The federal law still trails far behind California, Colorado, and New York in the ambition of its actions.

But in 2014, President Obama issued an executive order called the Climate Action Plan. Already, in Copenhagen in 2009, he had pledged that the country "would reduce its greenhouse gases emissions in the range of seventeen percent below 2005 levels by 2020." Now he was ordering the Environmental Protection Agency, under the Clean Air Act, to issue strict limits on power plants along with an array of other sources of pollutants, including particulates. In June 2014, the EPA released its state-specific, rate-based guidelines for existing power plants across the country.

New York State as a whole is well ahead of the curve in meeting these guidelines, but the investor-owned utilities are going to have to increase their efforts significantly to reach the reasonable goals set by this federal requirement. The Power Authority, thanks to its long-standing commitment to best practices, is already well placed to exceed the Federal requirements.

CHAPTER TWELVE

THE MEANING OF SANDY

Climate Change 101

The fundamental science of climate change is simple enough to demonstrate in any laboratory or greenhouse. Certain gases trap heat in the atmosphere, just like the glass panes of a greenhouse. That much is incontrovertible. Carbon dioxide is by far the most plentiful of the greenhouse gases, followed by methane (which traps thirty times more heat than carbon dioxide), then synthetic sulfur hexafluoride (SF_6), and a handful of other rare gases.

Since before the dawn of civilization, the composition of the Earth's atmosphere (78 percent nitrogen and 21 percent oxygen) has included carbon dioxide in a small but critical and consistent proportion of 278 parts per million (ppm). We know that fact from the traces of atmosphere trapped in core samples in ice, glaciers, and rock formations around the world, including the Arctic and Antarctica. That figure of 278 ppm remained stable for some 50,000 years until the Industrial Revolution began in the early 1800s, when the burning of coal increased exponentially and the level of carbon dioxide began rising.

At about that time, the worldwide human population crept up to one billion after remaining at a steady 200 million for at least two thousand years. It took another 130 years for the global population to arrive at two billion in 1930; but just thirty-two years later it reached three billion, and fifteen years later climbed to four billion in 1977. In 2016 the global population is nearing seven and a half billion, nearly forty times as many individuals as the Earth had supported since long before the pyramids. Almost all of those people aspire to have a home with modern conveniences, an automobile, and a modern place to work, though far too many aspire just to clean drinking water and a steady diet. Nevertheless, one must remember that it is overpopulation that drives the rising carbon dioxide in the atmosphere.

By 1958, when the National Oceanic and Atmospheric Administration (NOAA) began continuous monitoring of atmospheric carbon dioxide at the Mauna Loa Observatory in Hawaii, the level had risen from 278 to 315 parts per million. Less than 60 years later, on May 10, 2013, the carbon dioxide level in the Earth's atmosphere surpassed 400 parts per million for the first time in 800,000 years. Even the *rate* of increase is rising, now close to 3 ppm per year. Atmospheric methane has doubled since preindustrial times.

According to both NOAA's Global Analysis and the National Aeronautics and Space Administration (NASA), in 2014 the world experienced the warmest August since records began in 1880—coincidentally, the same year Thomas Edison devised the first coal-fired generating plant. Also since 1880, the average temperature of the entire planet has risen 1.53°F (.85°C); and NOAA reports that, globally, fifteen of the sixteen warmest years on record have occurred during the 21st century.

The Global Carbon Project, whose mission is "to develop a complete picture of the global carbon cycle," measured global CO_2 emissions due to fossil fuel use and cement production at 9.9 billion tonnes in 2013. That was 61 percent higher than in 1990 and 2.3 percent higher than the previous year. China's emissions accounted for 28 percent of the 2013 total; the United States for 14 percent; the European Union, 10 percent; and India, 7 percent. Only the European Union's twenty-eight states achieved a decline (-1.8 percent). Globally, CO_2 emissions (just from fossil fuel and cement production) come from: coal (43 percent), oil (33 percent), gas (18 percent), cement (5.5 percent), and gas flaring (0.6 percent).

This CO_2 amasses not only in the atmosphere but also in the oceans and on the land, where it is stored for varying periods of time in "carbon sinks," though most of these gases are eventually released back into the atmosphere. So some carbon sinks can become carbon sources, and others, like the oceans, can reach a critical point at which they cease to absorb carbon dioxide. There is some scientific quibbling about the time frames of such phenomena, but no fundamental disagreement. And it is established that each molecule of carbon dioxide endures for more than a hundred years. According to University of Chicago oceanographer David Archer in his book *The Long Thaw*, "The lifetime of fossil fuel CO_2 in the atmosphere is a few centuries, plus twenty-five percent that lasts essentially forever." And half the greenhouse gases in the atmosphere have been produced in recent decades, the result of the ceaseless rise in population; increased generation of electricity in India, China, and third-world countries; and the decimation and incineration of forests and rain forests.

Although globally one-third of all greenhouse gases comes from the generation of electricity, in the United States that fraction is more than double. According to the Environmental Protection Agency, in 2010 almost three-quarters of all US greenhouse gas emissions were released by power plants (excluding, of course, hydropower, solar, wind, etc.), the largest stationary source of carbon pollution in the United States. That is what makes the Power Authority's efforts so important and, in fact, urgent.

The Effects of Climate Change

Heat is energy, the excitation of molecules, and so the warmer the Earth's atmosphere becomes, the more dynamic. Much of the heat-trapping effect of greenhouse gases is transferred to the oceans, where it cycles in a predictable, annual circuit to Antarctica; in some areas, like the southwest Pacific Ocean north of Australia, the heat gathers in the ocean, and every few years it crosses eastward toward South America, thus producing the warmer "El Niño" effect shortly before Christmas in parts of the Americas. As well, all the land masses retain and radiate more heat, which further warms the atmosphere. As the heat melts more of the permafrost and glaciers, their white, reflective surfaces give way to darker geological bedrock that absorbs heat instead of reflecting it; and so in some places, widespread, ancient vegetation is uncovered, releasing continual methane emissions far worse than carbon dioxide. These and similar phenomena are the servo-mechanisms whereby global warming can accelerate in largely unpredictable ways. There is no way to anticipate all the ricocheting effects of the rising global temperature: even species can go extinct, altering ecosystems with unpredictable impact.

And while climate change cannot be blamed for any specific storm, the fact remains that carbon dioxide traps *energy* in the atmosphere in the form of heat, especially after the carbon dioxide level rises above 400 ppm. The first 100,000 feet above the Earth's surface—the troposphere, containing 80 percent of the atmosphere's mass—has become progressively more dynamic, and so seasonal air masses like the "polar vortex" have also shifted in unexpected ways. If serious scientists (i.e., *not* financed by the petroleum industry) were to deny humanity's role in changing the climate, they would need to explain why the annual release of billions of tons of such gases for centuries has *not* had an impact. No such explanation has ever been advanced with peer-reviewed research. So, when a series of unprecedented, monster storms repeatedly strike one region, it is disingenuous to insist that they have nothing to do with anthropogenic (man-made) gases released.

New York Superstorms

In late August 2011, a tropical cyclone that moved north from the Caribbean became Hurricane Irene and tore up the east coast from North Carolina, causing fifty-six deaths and ranking as the seventh most costly hurricane in US history, leaving an estimated $15.6 billion in damage.

Although Irene was downgraded to a tropical storm as it approached, Governor Cuomo anticipated the devastation by declaring a state of emergency two days before it struck New York on August 28, and by seeking a federal declaration from President Obama, which triggered the release of disaster assistance from the Federal Emergency Management Agency (FEMA). Cuomo was then able to order the deployment of 2,000 National Guard members to help police and transit workers cope with response and recovery. The Metropolitan Transportation Authority in the New York City area stopped all subway, bus, and rail service, and a handful of hospitals and nursing homes were evacuated.

Tropical Storm Irene brought flooding from the Hudson River into the Meatpacking District of Lower Manhattan, and throughout the construction site at the scene of the World Trade Center attack on September 11, 2001. Central Park recorded almost seven inches of rain and the South Shore of Long Island was inundated. There were more than 50 reports of sewage spills up the Hudson Valley, but the worst damage was further north in the Catskill Mountains, where the Schoharie Creek destroyed a third of the houses and businesses in the town of Schoharie, very near the Power Authority's Blenheim-Gilboa Pumped Storage plant. Still more damage from flooding and landslides ensued in the Adirondacks, as well as $45 million in crop damage. In all, some 600,000 New York homes and businesses lost power, more than half of those on Long Island. None of the statistics adequately reflect the personal loss and hardship that was inflicted as swaths of New York State faced months and even years of rebuilding.

Less than two weeks later, Tropical Storm Lee arrived from the Gulf. The enormous system moved very slowly, bringing less wind but far more rain to the west of Irene's path. Binghamton suffered record flooding and the downtown area was closed off as the Susquehanna River rose seventeen feet above flood stage, prompting the evacuation of 20,000 residents in Broome County. The village of Owego was 95 percent under water, and Interstate 88 was closed by a mudslide. The state suffered a further estimated $1 billion in destruction from this second storm.

For New York, the combination of Irene and Lee was unprecedented in its destructiveness, and left some regions reeling just five years after recovering from other massive storms. In each case, considerable preparations were taken around Manhattan, but the city did not get the full brunt of either storm, and though the subways were seriously threatened, they were not inundated.

A little over a year later, in October 2012, Hurricane Sandy formed as a tropical depression in the western Caribbean, and over the next 10 days became the second most destructive storm in US history (after Katrina), affecting twenty-four states, thirteen of which declared emergencies. It was vast, with a wind diameter reaching 1,150 miles and a record low barometric pressure upon landfall of 940 millibars. As the storm curved inland, its eye made a direct hit on Atlantic City. Sandy caused some $36 billion in damage in New Jersey and resulted in power outages for half of its customers that lasted for days and even weeks. Nationally, 286 people were killed, eleven million commuters had no service, and more than six million customers lost power (though some estimates range up to eight million), according to the US Department of Energy.

Then, on October 29, this meteorological behemoth pivoted north toward New York. By that time Obama had again responded to Cuomo's request, declaring in advance a federal emergency for New York, and the Governor had mobilized the National Guard, while all public transport was suspended. Even the stock exchanges were shut down for two trading days.

When "Superstorm Sandy" struck New York City, it was no longer classified as a hurricane, but that belied its impact, largely because of the wind-driven storm surge, which reached a record fourteen feet in Battery Park at about the time that a buoy in the harbor registered a wave 32.5 feet high. Six subway stops in Lower Manhattan were exposed to flooding through 540 openings, including the South Ferry Station, which had been rebuilt just three years earlier for half a billion dollars and was now destroyed under twenty feet of water. Then came an explosion at a Con Ed substation that knocked out all power to most of Lower Manhattan.

Elevators were frozen in place, leaving thousands to climb to their apartments. Many landlines and cell phone towers were disabled as hundreds of basements flooded, disabling even emergency generators installed there. Engineers at Bellevue Hospital, established in 1736, had had the foresight to place its emergency generators on the thirteenth floor; but when its basement filled with eighteen feet of water, the fuel pumps for those generators were submerged (along with essential computers, bulk oxygen tanks, and water pumps). So as the staff began evacuating patients, carrying them on stretchers down stairways, they were also carrying five-gallon jerrycans of fuel up thirteen stories. Four other city hospitals also had to be evacuated.

After the Governor visited the construction site at the World Trade Center, he described his descent seventeen stories below the surface to inspect a vast holding wall of welded iron, built to keep the Hudson River from flowing through every basement on the Lower West Side. "We were walking around with flashlights," Cuomo said later, "with water flowing from every direction." He called it one of the most extraordinary experiences in his life, looking up at that wall and realizing that nothing else was preventing the complete destruction of Lower Manhattan. "The whole southern tip of Manhattan was under water," he said. "It could not have taken much more of the flow that we were getting last night," which flooded miles of tunnels connecting New York and New Jersey.

Long Island bore the brunt of the residential wreckage. According to FEMA, Sandy damaged or destroyed 95,534 buildings in Nassau and Suffolk counties, leaving 4.4 million cubic yards of debris. These included forty-four power substations and other electric facilities, twenty-eight fire stations, twenty-six schools, and a medical facility. In adjoining Queens, more than a hundred homes in Breezy Point burned to the ground. Across the state an estimated 250,000 vehicles were destroyed. Staten Island's coast was also hit especially hard with a sixteen-foot high tide that destroyed hundreds of homes, impacting some 76,000 residents and killing twenty-three.

One could fill volumes recounting the heroic efforts carried out by ordinary folk in New York and the myriad tales of individual tragedy, especially among the poor and infirm. Despite gargantuan help and cooperation from the federal government for both New York and New Jersey, in some areas of Lower Manhattan, the Rockaways, Long Island, Westchester, and points north, the effort needed to recover seemed incalculable and unachievable. For just over two million New Yorkers without power for a week or more, daily survival was the daunting, immediate challenge. By good fortune, the Indian Point 2 nuclear plant continued operating at 100 percent power; Indian Point 3 was unharmed, but had to be shut down because of other issues with the grid.

The secondary impact of Superstorm Sandy began as gas stations closed across the mid-Atlantic and Northeast, since the stations could not open without electric power. But that was only part of the problem. The wholesalers that filled the tanker trucks were also without power, which meant that they could neither receive nor provide fuel. Further up the "gas chain," several refineries in New Jersey and beyond were shut down, and even the equipment that transferred crude oil from ships to the refineries was useless. Without gas, recovery was slowed significantly as scores of cars waited in line at just a handful of gas stations; so did many of the power companies' repair vehicles. Ten days after Sandy struck, Mayor Michael Bloomberg imposed gas rationing in New York City: drivers with even-numbered license plates could buy gas on even-numbered dates, those with odd plates on odd dates. The rationing continued for another two weeks.

Another, smaller storm in the region added insult to injury, as it were, ten days after Sandy, but this time several inches of snow and ice fell upon the dark, desolate, damaged homes, further hampering the transportation of people, food, and water. There were even additional power outages. Many families and businesses in New York that were just recovering from the financial crisis of 2008 and the ongoing unemployment crisis were dealt further crushing blows by the massive storms of 2011 and 2012.

It is a painful irony that such long, widespread power outages were the result of climate change caused in large part by profligate power production over the decades, since one-third of all greenhouse gases worldwide are emitted by the power industry. Instead of merely making feeble efforts to adjust to a "new reality" that is changing faster than we are, it behooves us all to acknowledge reality and take action, individually and collectively. We have no other rational choice.

Developing a Clean Energy Economy

In its aftermath, the meaning of Sandy was clear. New York's governor wasted no time in articulating it. Less than three weeks after the storm struck, Cuomo published an article in the *New York Daily News* under a headline that prioritized the themes of addressing global warming *and* preparing for its impact. "We Will Lead on Climate Change—N.Y. must press ahead with urgency to equip itself for the new age of extreme weather." In the article, Cuomo wrote,

> Extreme weather is the new normal. In the past two years, we have had two storms, each with the odds of a 100-year occurrence. Debating why does not lead to solutions—it leads to gridlock.... Recent events demand that we get serious once and for all. We need to act, not simply react.
>
> Our electrical power grid and the structures that control it must undergo a fundamental redesign. Power utilities are the equivalent of vinyl records in the age of the iPod: antiquated, 1950s-style institutions that don't serve our current needs. The electrical system is particularly vulnerable. Entire above-ground wire networks in heavily wooded areas spell disaster in virtually every storm, yet nothing is being done to rectify this obvious vulnerability. To a large degree, the state and local governments are captive to the utilities in an emergency, just like their customers. Six utilities do not compete for customers, but in a natural disaster they must compete against one another for scarce materials and personnel. We must investigate the weaknesses in the system and reform it to ensure that customers do not face catastrophic power losses every few years.... There is no more time for debate. This is our moment to

> act.... We will not allow the national paralysis over climate change to stop us from pursuing the necessary path for the future.

In the following weeks, the governor announced the formation of four separate commissions that would address "how we get ready before an event, how we respond in its immediate aftermath, and changes we can make to our infrastructure that will better prepare us to face Mother Nature's inevitable fury." As a result of these commissions, many of the utilities' problems revealed by Sandy were addressed by improving mutual assistance between them, flood preparations, power restoration times, communication with customers, coordination with local governments, and specific training for the National Guard.

Even before the superstorm, Cuomo had taken aim at a higher goal than clean energy and economic development. Since taking office, he had sought to foster a *clean energy economy* that generates more power and more jobs while reducing gas emissions by supporting all the possible advancements: solar, wind, biomass, oil-to-gas conversion, electric vehicles, and, of course, energy efficiency. This last option is perhaps the most promising for now, because so much of the more than 220 million square feet of real estate owned by the state sorely needs effective insulation, heating boilers, efficient lighting, etc., which is considered "low-hanging fruit" in energy conservation. State properties produce fully half the state's greenhouse gases from generation, so if Executive Order 88 and the BuildSmart NY program succeed in reducing those by 20 percent before 2020, that effort alone may reduce the state's total emissions very significantly in just seven years.

The lessons learned from Sandy raised the priority of still other considerations such as "hardening" the entire power infrastructure, especially transmission lines and vulnerable substations, to withstand hurricanes in the future. Implementation of these measures would have to be integrated with long-range development of a "Smart Grid" that can forecast customer usage and also accommodate the addition of smaller, variable generators (e.g., wind and rooftop solar photovoltaic) feeding into the larger grid.

Smart Grid offers many benefits: first, in a power failure, all the utilities and the state's Independent System Operator receive instantaneous details about all affected locations by way of smart meters at customer sites, providing a new level of reliability. The data relayed by synchrophasor technology also reduces the need for generating capacity and facilitates all sorts of efficiency measures. California leads the large states in Smart Grid technology that increases participation in the wholesale market by allowing for utility-scale renewable resources, incentive-based demand response, energy storage, and small-scale solar photovoltaics.

In October 2014, New York took a crucial step in that direction by launching the New York Energy Manager, developed, deployed, and managed by the Power Authority. This facility was built within the College of Nanoscale Science and Engineering at SUNY Polytechnic Institute in Albany to serve, Cuomo said, "as a statewide energy monitoring hub" that "leverages the very best in smart technology to help the State save millions of dollars a year on energy costs." Although by now there are existing Smart Grid systems that can be studied and compared, considerable research and cost/benefit analysis remains to be done for the particular features of transmission and distribution in New York State. In the end, the architecture of this transformation will be determined by the Public Service Commission, the major investor-owned utilities, the municipal utilities, and NYPA. But its long-term value will depend on its ability to accommodate and encourage the development of more renewable energy resources.

Reforming the Energy Vision

The Cuomo administration has increased its many solar installation programs, including K-Solar (with NYSERDA), which serves schools statewide. In 2013, the United States installed around 1GW on residential rooftops, but by 2016 the installation rate is expected to rise sixfold. As of 2014, according to one energy analyst, in ten US states the cost of installing rooftop photovoltaic (PV) systems was the same *or less* than that for power purchased from the grid, and by 2016 the same "grid parity" will be achieved by solar rooftop in all fifty states, including New York. NYPA is even testing solar PV carports designed and installed solely to charge electric vehicles. Small windfarms are being evaluated in different locations, while plans are presently unfolding for a very large offshore wind turbine array 10 miles south of Long Island that is inching its way through the regulatory hoops. There are smaller efforts for harnessing wind inland, as well as some biomass units being developed in the farmlands.

But with the lessons from Sandy still fresh, it gradually dawned on many officials that addressing *separately* such distinct issues as resilience, Smart Grid, renewable energy, security, and efficiency might produce unintended conflicts among them. For example, would the expensive and sensitive equipment needed for a Smart Grid be even more prone to damage in a hurricane? Would the electronic communications at the heart of a Smart Grid make it more vulnerable to cyberattack? Could adding thousands of small photovoltaic units pose a security threat to the grid, statewide and beyond? Could resiliency efforts taken right away prove obsolete with the development of Smart Grid? But over all these questions loomed a larger issue: would New York's six investor-owned power companies join with the state in a genuine collaboration that might involve reducing their profits?

The Public Service Commission, in exactly the spirit that Governor Hughes had expressed in 1907 and that Governor Cuomo encouraged in 2014, announced its intention to take a broader perspective on all the issues at hand in a major initiative entitled *Reforming the Energy Vision*. Its report began by looking back at how the New York system had evolved from Edison, Westinghouse, and Tesla. "For most of its history, the basic design of the electric grid has remained essentially the same," the report said. "Electricity is generated at central stations, transmitted long distances via high-voltage lines, then stepped down in voltage and delivered to customers through local distribution systems. When the system was developed, this design was needed in part because of the limitations of pre-computer-age communications, and in part because customer demand was relatively inelastic."

Reforming the Energy Vision does not put forward final solutions, but rather calls for "a fundamental reconsideration of our regulatory paradigms and markets, examining how policy objectives are served both by clean energy programs and by the regulation of distribution utilities." The commissioners are seeking to affect a profound revolution in power generation, transmission, and distribution in the state, a very different model from the coal-based central plant introduced by Thomas Edison 130 years ago on Pearl Street. In large part, they are aiming to redefine "the role of the distribution utilities in enabling system wide efficiency and market-based deployment of distributed energy resources and load management," and to make changes in the "current regulatory, tariff, and market design and incentive structures in New York to better align utility interests with achieving our energy policy objectives." The new system must allow for a greater diversity and decentralization in renewable generation, and also provide customers with tools that can offer them better management of their total energy bill.

Decentralization is key to this transformative vision for the entire New York energy structure. For the Power Authority, a "sea change" is coming, according to President and CEO Gil Quiniones, who describes the process of reinventing NYPA as a new, more distributed grid takes shape. "The investments that we will make, and the technology choices that we will decide over the next ten years, will determine what NYPA and what the grid will look like over the next forty years," he says. "Our customers have been clamoring to have more control over how they use the energy that we sell to them.... We've done energy efficiency in the past" in part because NYPA was mandated to do so, but "now it is really more customer driven. Energy efficiency, demand response, renewables, solar, combined heat and power, microgrids—you are going to see NYPA do more and more of those."

Looking to the Future

In 2015 the Power Authority had seven microgrid projects under consideration or in development, including a 15-MW gas-fired Cogeneration

Heating Plant (CHP) system on Riker's Island, where the prison population numbers some 15,000 inmates and workers. "If there is another blackout in New York," Quiniones said, "Riker's Island will be up and running." The Authority is also assessing the feasibility of a microgrid for the New York City Housing Authority's 2,800 apartments in Red Hook, Brooklyn, an especially vulnerable waterfront location. The study will analyze local energy sources, including CHP, emergency generators, wind, or solar energy. Many of the Housing Authority's other developments might also be suited for microgrid development, and five other microgrids are under consideration, upstate and down. "We're really changing the rules of the road and telling utilities that they should modernize their grid to foster the connections of all of these distributed resources," Quiniones noted. "It's almost like [the utilities] are the iPad and what they need to do is to encourage the private energy service companies to develop various apps that the iPad can use. From NYPA's perspective, for our set of customers we want to help develop those apps, so that we can help our customers in this new environment."

This new environment represents the first fundamental redesign of the central-plant model that Thomas Edison had created in order to sell his light bulbs. In fact, a system that includes rooftop solar panels bears a closer resemblance to the private generators Edison first installed in the home of J. P. Morgan. As a monopolist, Morgan himself preferred the model of private generation because it would encourage the sale of more generators while reserving the benefits of electric lighting and power for the wealthy. But the dependence upon coal, along with other market forces, determined that central plants would win the day, circa 1900. And once the investor-owned utilities had developed intraregional monopolies, their interests were better served by the status quo than by any innovations to the grid. Other potential systems, such as small-scale hydropower, were rarely considered, and the course for the future was set.

"In the years between 1890 and 1920," wrote historian David E. Nye, "a dialogue on electricity took place within American society.... Americans had to choose whether to construct many small generating stations or a centralized system, whether to use alternating current or direct current, whether to place ownership in public or private hands, whether to establish rates that favored the small or the large consumer, whether to give control over the system to technicians or to capitalists or to politicians." Once those crucial decisions had been made, the course was clear and the addition of state-owned power further strengthened the centralized system, providing an unacknowledged, underlying guarantor of reliable service, especially in extreme circumstances.

A new architecture is evolving for the New York grid in the twenty-first century, along a path that embodies Governor Cuomo's plans for "Reforming the Vision" of generated electricity statewide, as well as the Power Authority's "Strategic Vision." Along with NYSERDA and the NYISO, NYPA remains central to innovation and to the state's economic foundation.

It is difficult to imagine just how the state of New York would have evolved in the past century if the utilities that were privately owned—mostly by Morgan—had gone unchecked by the Public Utilities Commission and had added hydropower from the St. Lawrence and Niagara rivers to their monopoly. One can only say with certainty that the state would have developed very slowly and differently if each governor since Hughes had not fought for the principle that access to electricity is a necessity for every modern business and residence. Governor Cuomo has gone much further. "New York," he declares firmly, "is on its way to reclaiming its place as a model for the nation and the world. We must not turn back now."

Notes

CHAPTER ONE

1 **Toward that end.** Cited in *Bold Dream... Shining Legacy, Seventy-Fifth Anniversary of NYPA History* (NYPA, 2006).

2 **To accomplish this.** *Public Power in America: A History* (American Public Power Association), p. 1.

9 **As one visitor wrote.** Jill Jonnes, *Empires of Light: Edison, Tesla, Westinghouse and the Race to Electrify The World* (Random House, 2003), p. 47.

10 **Jill Jonnes's *Empires of Light*.** Jonnes, *op. cit.*

10 **"Anything that won't sell.** Cited in Harold Evans, *They Made America: From the Steam Engine to the Search Engine, Two Centuries of Innovators* (Back Bay Books, 2004), p. 180.

11 **"He picked an area.** Leonard S. Hyman, *America's Electric Utilities: Past, Present and Future* (Public Utilities Reports, Inc., Arlington, VA, 1997), p. 85.

11 **It generated enough power.** Richard Rudolph and Scott Ridley, *Power Struggle: The Hundred-Year War Over Electricity* (New York: Harper Collins, 1986), p. 28.

11 **According to the article.** Leonard Franklin Page, *Evolution of the Electric Incandescent Lamp* (H. Cook, 1889), pp. 38–9.

12 **Despite Edison's best arguments.** Hyman, *op. cit.*, p. 8.

12 **Edison's biographer Paul Israel.** Paul Israel, *Edison: A Life of Invention* (Wiley and Sons, 1998), p. 167.

12 **As Jonnes explained.** Jonnes, *op. cit.*, p. 67.

13 **New York already had.** David Nye, *Electrifying America: Social Meanings of a New Technology* (MIT Press, 1990), p. 47.

13 **By this time.** "Moonlight on Broadway," *New York Evening Post,* Dec. 21, 1880.

13 **New York City's electrification.** Thomas P. Hughes, *Networks of Power: Electrification in Western Society, 1880–1930* (Baltimore: Johns Hopkins University Press, 1983), p. 227.

14 **Hyman wrote that Edison's.** Hyman, *op. cit.*, p. 85.

CHAPTER TWO

16 **"I drew with a stick.** John J. O'Neill, *Prodigal Genius: The Life of Nikola Tesla* (New York: McKay, 1944), p. 49, cited in Jill Jonnes, *Empires of Light: Edison, Tesla, Westinghouse and the Race to Electrify The World* (Random House, 2003), p. 93.

16 **As a journalist.** Arthur Brisbane, [*New York*] *Sunday World*, July 22, 1894, p. 5.

16 **Jonnes explains that.** John J. O'Neill, *op. cit.*, p. 49, cited in Jonnes, *op. cit.*, pp. 94–5.

17 **Tesla later critiqued.** Cited in Margaret Cheney, *Tesla: Man Out of Time* (Simon and Schuster, 2011), p. 55.

17 **According to Tesla.** O'Neill, *op. cit.*, p. 64.

17 **Edison, who had been.** Cited in Tom McNichol, *AC/DC: The Savage Tale of the First Standards War* (John Wiley and Sons, 2011), p. 43.

17 **Tesla said that when.** O'Neill, *op. cit.*, p. 62.

17 **Two years after Pearl Street.** Leonard S. Hyman, *America's Electric Utilities: Past, Present and Future* (Public Utilities Reports, Inc., Arlington, VA, 1997), p. 85.

18 **As for the large.** Forrest McDonald, *Insull* (Chicago: University of Chicago Press, 1962), p. 26.

18 **Still, "the Edison system,"** Thomas P. Hughes, "Technological History and Technical Problems," in Chauncey Starr and Philip C. Ritterbush, eds., *Science, Technology and the Human Prospect* (New York: Pergamon Press, 1980), p. 144.

20 **As salesmen from.** *Public Power in America: A History* (Washington, D.C.: American Public Power Assn.), p. 5.

21 **"Just as certain as death,"** Matthew Josephson, *Edison: A Biography* (New York: McGraw-Hill, 1959), p. 346.

21 **And Edison, boasted.** "A Warning from the Edison Electric Light Co.," Feb. 1888, p. 45, retrieved from archive.org/stream/warningfromediso00edis#page/n33/mode/2up [Retrieved 4/3/14].

21 **Gradually Edison was.** Report ... Transmitted to the Legislature of the State of New York, Jan. 17, 1888 (Argus, 1888), p. 80.

22 **Another ardent opponent.** Cited in Mark Essig, *Edison and the Electric Chair: A Story of Light and Death* (Bloomsbury, 2006), p. 141.

22 **When his attorney.** "Testimony of the Wizard: Edison's Belief in Electricity's Fatal Force," *New York Times*, July 24, 1889, p. 2.

22 **After the grisly.** Maury Klein, *The Power Makers: Steam, Electricity, and the Men Who Invented Modern America* (Bloomsbury, 2010), p. 178.

23 **As several witnesses.** "Far Worse than Hanging: Kemmler's Death Proves an Awful Spectacle," *New York Times*, Aug. 7, 1890, p. 1.

24 **"Of all the forms.** Marc J. Seifer, *Wizard: The Life and Times of Nikola Tesla* (New York: Citadel, 1996), p. 71.

25 **Biographer Henry Prout wrote.** Henry G. Prout, *A Life of George Westinghouse* (New York: Scribner's, 1926), p. 293.

25 **"Though past forty.** Nikola Tesla, "Tribute to George Westinghouse," *Electrical World and Engineer*, March 21, 1914, p. 637, cited in Seifer, *op. cit.*, p. 52.

25 **"A dollar given.** Francis Ellington Leupp, *George Westinghouse: His Life and Achievements* (Little, Brown, 1918), p. 294.

26 **This was perfectly.** John J. O'Neill, *op. cit.*, p. 49.

26 **Surely, wrote the editor.** "Mr. Edison's Mistake," *Electrical Engineer*, Feb. 17, 1892, p. 162.

27 **Finally, the Westinghouse men.** Seifer, *op. cit.*, p. 60.

28 **There were almost 60,000.** *Ibid.*, pp. 118–9.

29 **"My experience over there.** Josephson, *op. cit.*, p. 429.

CHAPTER THREE

31 **According to ecologist.** Polly Higgins, "The Origins of Hydroelectricity," *The Ecologist*, Sept. 6, 2007.

31 **Armstrong also planted.** *Ibid.*

32 **"It would be hard.** Charles Dickens to John Forster, April 26, 1842, *Life, Letters and Speeches of Charles Dickens* (New York: Houghton, Mifflin, 1894), Vol. 2, p. 168.

32 **When the Duke.** Duke de la Rochefoucauld, *Travels Through the United States of North America....*," cited in the *Anthology and Bibliography of Niagara Falls*, 2 vols., Charles Mason Dow, ed. (Albany: State of New York), II, p. 939.

32 **Brush installed.** "Niagara Falls History of Power Development," www.niagarafallsinfo.com%2Fhistory-item.php%3Fentry_id%3D1435%26current_category_id%3D242&ei=W4NrU_uBOuXlsATsm4C4CQ&usg=AFQjCNHh_NOHU3R4mVkFBBfRlOec8u6RnA&sig2=7sKA-1WRQ2VZbOP8OcAP0w [Retrieved 4/25/14].

33 **A Rochester engineer.** Pierre Berton, *Niagara: A History of the Falls* (New York: Penguin, 1992), p. 207.

33 **Although 200 hydroelectric.** Dept. of Energy, "History of Hydropower," www.energy.gov/eere/water/history-hydropower [Retrieved 5/2/14].

33 **Also largely untested.** Edward Dean Adams, *Niagara Power*, Vol. 2, (Niagara Falls: Niagara Falls, Power Co., 1927), p. 173.

34 **Nevertheless, Morgan and Adams.** George Forbes cited in "Electric Launches at Chicago," *The Electrician*, Nov. 17, 1893, in *The Electrical Journal*, Vol. 32, D. B. Adams, ed., p. 70.

34 **"America is in no.** Lewis Buckley Stillwell, "Electric Power Generation at Niagara," cited in the *Anthology and Bibliography of Niagara Falls*, Vol. 2, Charles Mason Dow, ed. (Albany: State of New York), pp. 941–3.

34 **This was the golden.** Cliff Spieler and Tom Hewitt, *Niagara Power from Joncaire to Moses* (Lewiston, NY: Niagara Power Publishers, 1959), p. 12.

34 **"Along the edge.** George Forbes, "The Utilization of Niagara," *Electrical Engineer*, Vol. 15, Jan. 18, 1893, p. 65.

36 **As well, many new.** Berton, *op. cit.*, p. 170.

37 **"The current sent.** "Niagara is Finally Harnassed," *New York Times*, Aug. 27, 1895, p. 9.

37 **These dynamos and turbines.** H. G. Wells, "The End of Niagara," *The Future in America: A Search After Realities* (Harper and Bros., 1906) p. 74 [emphasis added].

37 **Of the hydropower.** Wells, *op. cit.*, p. 75.

38 **As Jill Jonnes elegantly.** Jill Jonnes, *Empires of Light: Edison, Tesla, Westinghouse and the Race to Electrify The World* (Random House, 2003), p. 329.

38 **"Electrical experts say.** *Buffalo Enquirer*, "Yoked to the Cataract!," Nov. 16, 1896, p. 1, cited in Jane Brox, *Brilliant: The Evolution of Artificial Light* (Houghton Mifflin Harcourt, 2010), p. 149.

38 **And of course, unlike.** Rank by Population of the 100 Largest Urban Places, Listed Alphabetically by State: 1790–1990, US Census Bureau [Retrieved 5/2/14].

39 **Lighting "drew more trade.** David Nye, *Electrifying America: Social Meanings of a New Technology* (MIT Press, 1990), pp. 53–4.

39 **Nye also noted that.** Nye, *op. cit.*, p. 15.

41 **According to renowned climatologist.** James Hansen *et al.*, "Target Atmospheric CO_2: Where Should Humanity Aim?" *Open Atmospheric Science Journal* (2008), Vol. 2, pp. 217–231 [arXiv:0804.1126, p. 12; Retrieved 5/28/14].

42 **Two years later.** Ernest Heinrichs cited in Quentin R. Skrabec, *George Westinghouse: Gentle Genius* (Algora Publishing, 2006), p. 235.

42 **"Their very magnitude,"** *Ibid.*, p. 240.

CHAPTER FOUR

44 **To understand the helplessness.** Robert McChesney and John Podesta, "Let There Be Wi-Fi," *Washington Monthly*, Jan./Feb. 2006.

44 **In response, private utility.** *Ibid.*

45 **They descended from.** Richard Zacks, *Island of Vice: Theodore Roosevelt's Doomed Quest to Clean Up Sin-Loving New York* (New York: Doubleday, 2012), p. 19.

45 **He described "going down.** *Ibid.*, p. 21.

46 **That was because although.** Kenneth D. Ackerman, *Boss Tweed: The Rise and Fall of the Corrupt Pol Who Conceived the Soul of Modern New York* (New York: Carroll & Graf, 2005).

46 **Tweed's system had.** *Ibid.*, p. 277.

47 **But mutual cooperation.** Herbert Mitgang, *The Man Who Rode The Tiger: The Life and Times of Judge Samuel Seabury* (Philadelphia: Lippincott, 1963), p. 32, cited in Robert E. Wesser, *Charles Evans Hughes: Politics and Reform in New York, 1905–1910* (Ithaca: Cornell University Press, 1967), p. 72.

47 **As one leader.** Robert E. Wesser, *Charles Evans Hughes: Politics and Reform in New York, 1905–1910* (Ithaca: Cornell University Press, 1967), p. 18.

48 **The future president's.** Zack's, *op. cit.*

48 **According to the New York Police.** New York Police Department website: web.archive.org/web/20060930063300/nyc.gov/html/nypd/html/3100/retro.html [Retrieved 7/20/14].

49 **The *New York World*.** Zacks, *op. cit.*, p. 121.

49 **Platt's political motto.** *Ibid.*, p. 160.

49 **As TR wrote.** *Ibid.*, p. 161.

50 **By 1905.** Wesser, *op. cit.*, p. 22.

51 **There were also more.** Richard Rudolph and Scott Ridley, *Power Struggle: The Hundred-Year War Over Electricity* (New York: Harper Collins, 1986), p. 38.

51 **Most of these were.** American Public Power Association, *Public Power in America: A History* (Washington: APPA, 1994), p. 5.

52 **He accused seventeen.** Rudolph and Ridley, *op. cit.*, p. 23.

52 **"I believe in municipal.** *Ibid.*, p. 22.

52 **The city fought J. P. Morgan.** *Ibid.*, p. 38.

52 **were the revelations.** Wesser, *op. cit.*, p. 30.

53 **It was President Roosevelt.** Roosevelt to Hughes, Oct. 5, 1906, *Letters of Theodore Roosevelt, V*, pp. 443–4.

53 **As Wesser wrote, Hughes.** Wesser, *op. cit.*, pp. 87–88.

53 **But Hughes held to his.** Barbara H. Brock, *The Development of Public Utility Accounting in New York* (East Lansing, Michigan: Michigan State University, 1981), p. 5.

53 **He consistently hewed.** Wesser, *op. cit.*, p. 103.

53 **As for the Republicans'.** Frank E. Kilroe, "The Governorship of Charles Evans Hughes: A Study in Reform (1906–1910)" (unpublished M.A. Thesis, Columbia University, 1934), p. 24.

54 **Hughes was elected.** *Ibid.*

54 **"My feelings,"** *New York Times*, Nov. 7, 1906.

54 **The same day.** Roosevelt to Hughes, Nov. 7, 1906, *Letters of Theodore Roosevelt, V*, pp. 490–1.

54 **Hughes's inaugural speech.** *New York Evening Post*, Jan 2, 1907; *New York Herald*, Jan. 3, 1907; *New York Times*, Jan. 3, 1907.

54 **Both commissions would.** Wesser, *op. cit.*, p. 155.

55 **A primary goal.** Barbara H. Brock, *The Development of Public Utility Accounting in New York* (East Lansing, Michigan: Michigan State University, 1981), p. 5.

55 **The bill aimed.** Hughes at Elmira, May 3, 1907, cited in Merlo J. Pusey, *Charles Evans Hughes* (New York: Macmillan, 1951), I, p. 203.

55 **Each commission would have.** *Public Papers of Charles E. Hughes, 1907* (New York: J. B. Lyon, 1909), pp. 31–2.

56 **I am here under.** *Ibid.*, p. 206.

56 **When one state senator.** *World's Work*, March, 1908, p. 41.

56 **That June, when the.** Pusey, *op. cit.*, p. 207.

56–57 **Hughes found that unacceptable.** *Public Papers of Charles E. Hughes, 1907* (New York: J. B. Lyon, 1909), p. 37 [emphasis added].

57 **That policy was based.** *Public Papers of Charles E. Hughes, 1910* (New York: J. B. Lyon, 1909), pp. 22–5, cited in Pusey, *op. cit.*, pp. 213–4.

CHAPTER FIVE

59 **Their subsequent conversations.** Ernest K. Lindley, *Franklin D. Roosevelt: A Career in Progressive Democracy* (Indianapolis: Bobbs-Merrill, 1931), p. 69.

60 **"Mr. Roosevelt carried.** Cited in *Ibid.*, p. 75.

60 **He was not yet.** *New York Times*, January 22, 1911.

60 **"From the ruins.** Lindley, *op. cit.*, p. 101.

61 **The technology that.** *Ibid.*, p. 159.

62 **Reformers like FDR.** Robert Caro, *The Power Broker: Robert Moses and the Fall of New York* (New York: Vintage, 1974), p. 94.

62 **Smith was still Tammany's.** Cited in Judson King, *The Conservation Fight: From Theodore Roosevelt to the Tennessee Valley Authority* (Washington, D.C.: Public Affairs, 1959), pp. 19–20.

63 **TR spoke in upstate.** Cited by Smith, *Denver Sunday Morning Star*, September 23, 1928, p. 4.

63 **Smith's waterpower policy.** *Bold Dream... Shining Legacy, Seventy-Fifth Anniversary of NYPA History* (NYPA, 2006).

64 **"Giant power combines.** *Ibid.*

65 **Beginning with his.** Lindley, *op. cit.*, p. 239.

65 **And he insisted.** *Ibid.*

65 **There remains the technical.** Cited in Lindley, *op. cit.*, pp. 239–40.

66 **While the states regulated.** Cited in Arthur M. Schlesinger, *The Age of Roosevelt: The Crisis of the Old Order* (Boston: Houghton Mifflin, 1957), p. 119.

66 **By 1932, according.** Leonard S. Hyman, *America's Electric Utilities: Past, Present and Future* (Public Utilities Reports, Inc., Arlington, VA, 1997), p. 106.

66 **"At the top of.** *Ibid.*, p. 10.

66 **In 1926 alone.** Richard Rudolph and Scott Ridley, *Power Struggle: The Hundred-Year War Over Electricity* (New York: Harper Collins, 1986), p. 50.

67 **Pennsylvania Governor Gifford Pinchot.** *Ibid*, p. 50.

67 **The bill to form.** *Bold Dream... op. cit.*

67 **"Throughout the summer.** Lindley, *op. cit.*, p. 250.

68 **Eschewing any highfalutin'.** *Bold Dream... op. cit.*

69 **Still, Insull insisted.** Samuel Insull in National Electric Lighting Association, Proceedings, 1898, p. 27.

69 **Insull had recognized.** Rudolph and Ridley, *op. cit.*, pp. 38–9.

69 **There was no law.** *Ibid.*, p. 52.

70 **As Hyman explains.** Hyman, *op. cit.*, p. 106.

70 **"By bloating the value.** Rudolph and Ridley, *op. cit.*, p. 53.

70 **As the US ambassador to Germany.** Lindley, *op. cit.*, pp. 317–8.

70 **Insull's personal empire.** Hyman, *op. cit.*, p. 106.

70 **One year after.** *Bold Dream... op. cit.*

71 **But "sensing that he held.** Lindley, *op cit.*, pp. 251–2.

71 **There was a new urgency.** *Ibid.*, p. 251.

71 **Perhaps that is why.** *Ibid.*

71 **However, rather than reintroduce.** Lindley, *op cit.*, p. 251.

71–72 **He promptly sent.** *Ibid.*

72 **"It is a milestone,"** *Bold Dream... op. cit.*

72 [He] **said that it was.** Lindley, *op cit.*, pp. 257–8.

73 **As biographer Lindley.** *Ibid.*, p. 263.

73 **In Binghamton.** *Bold Dream... op. cit.*

74 **Norris "exhibited photostat.** King, *op. cit.*, p. 128.

74 **When TR's Inland Waterways.** This is closely examined in Preston John Hubbard, *Origins of the TVA* (Nashville: Vanderbilt University, 1961).

75 **Hoover was especially wary.** Sarah Phillips, "FDR, Hoover, and the New Rural Conservation," in *FDR and the Environment*, Henry L. Henderson and David B. Woolner, eds. (New York: Palgrave Macmillan, 2009), p. 142.

76 **They had, wrote Lindley.** Lindley, *op cit.*, pp. 329–30.

77 **He also proposed that.** *Ibid.*

77 **This project would represent.** *Ibid.*

77 **"Within twenty-four hours.** *Ibid.*, p. 331.

77 **"Stronger than all these.** *Bold Dream... op. cit.*

78 **In a radio address.** *Ibid.*

78 **Instead, an instrumentality.** *Laws of New York, 1931*, p. 1645 (Chap. 772, April 27, 1931, p. 6).

78 **The Act also specifically.** *Ibid.*, p. 9. Clause 8 states that "No bonds or other obligations of the power authority shall be issued until firm contracts for the sale of power shall have been made by it sufficient to insure payment of all operating and maintenance expenses of the project, and interest on, and amortization and reserve charges sufficient to retire, the bonds power authority issued for the project in not more than fifty years from the date of issue thereof."

79 **"State owned or federal owned.** King, *op. cit.*, pp. 261–2.

CHAPTER SIX

82 **"Our great river.** Daniel Macfarlane, *Negotiating a River: Canada, the US, and the Creation of the St. Lawrence Seaway* (British Columbia: UBC Press, 2014), p. 30.

82 **"The United States was.** *Ibid.*, p. 46.

82 **Despite Bennett's personal distaste.** *Ibid.*, p. 35.

82 **So Bennett and Hoover.** *Department of State Publication No. 347*, Great Lakes–St. Lawrence Deep Waterway Treaty.

82 **However, as Macfarlane notes.** Macfarlane, *op. cit.*, p. 37.

83 **There were also serious.** Philip C. Jessup, "The Great Lakes–St. Lawrence Deep Waterway Treaty," *American Journal of International Law*, Vol. 26, No. 4, Oct. 1932, p. 817.

83 **Still, the most contentious.** *Ibid.*

83 **On his thirty-seventh day.** Judson King, *The Conservation Fight: From Theodore Roosevelt to the Tennessee Valley Authority* (Washington, D.C.: Public Affairs, 1959), p. 268.

84 **"His unique contribution.** *Ibid.*, p. 267.

84 **But beyond that.** *Ibid.*, p. 268.

85 **The TVA Act specified.** "The Tennessee Valley Authority, Dammed if You Don't," *The Economist*, April 27, 2013, www.economist.com /news/business/21576682-barack-obama-mulls-privatising-americas-biggest-public-utility-dammed-if-you-dont [Retrieved 2/19/15].

85 **Once again FDR, goaded.** King, *op. cit.*, p. 269.

85 **With that, wrote Judson.** *Ibid.*, pp. 275–6.

85 **FDR passed many.** *Bold Dream... Shining Legacy, Seventy-Fifth Anniversary of NYPA History* (NYPA, 2006).

86 **The BPA was initially.** *Ibid.*

86 **When Prime Minister King.** Macfarlane, *op. cit.*, p. 40.

87 **The Act, wrote power.** Leonard S. Hyman, *America's Electric Utilities: Past, Present and Future* (Public Utilities Reports, Inc., Arlington, VA, 1997), p. 115.

87 **"The fight over.** Rudolph and Ridley, *op. cit.*, p. 76. The Edison Electric Institute was established in 1933 and today represents all of the investor-owned utilities.

88 **The impact on.** Hyman, *op. cit.*, p. 116.

89 **After interprovincial agreement.** *NYPA Annual Report, 1941*, p. 10.

89 **Significantly, FDR continued.** *Bold Dream... op. cit.*

89 **But he was advised.** *Ibid.*

89 **Even New York supporters.** *Ibid.*

91 **The US steel industry.** Austin W. Clark, "Fish or Cut Bait? Dwight D. Eisenhower and the Creation of the St. Lawrence Seaway" (unpublished thesis, Gettysburg College, 2012), p. 7.

91 **The Truman administration.** Macfarlane, *op. cit.*, pp. 58, 60.

91 **Those delays were.** Clark, *op. cit.*, p. 5.

92 **NYPA's 1951 Annual Report.** *Bold Dream... op. cit.*

92 **The Toronto *Globe and Mail*.** Macfarlane, *op. cit.*, p. 67.

92 **"Going ahead with.** Chevrier, *The St. Lawrence Seaway* (Toronto: Macmillan), pp. 47–8.

92 **This was a bold.** www.statcan.gc.ca/pub/98-187-x/4151287-eng.htm [Retrieved 3/11/15].

93 **Public opinion supported.** Macfarlane, *op. cit.*, p. 71.

93 **Says Macfarlane, "The hope.** *Ibid.*, p. 72.

93 **Truman bluntly threatened.** R. R. Baxter, *Documents on the St. Lawrence Seaway* (New York: Frederick A. Praeger Inc., 1960), p. 19.

94 **"The onus for this project,"** Clark, *op. cit.*, pp. 9, 11.

94 **They were all in tune.** AAR pamphlet, cited in Clark, *op. cit.*, p. 13.

95 **One senator told.** Sen. Leverett Saltonstall, cited in Clark, *op. cit.*, pp. 13, 14.

96 **As chairman of the Joint.** Clark, *op. cit.*, p. 17.

96 **According to historian Clark.** *Ibid.*, p. 19.

96 **And, as Clark reports.** *Ibid.*, p. 27.

96 **As he later explained.** Dwight David Eisenhower, *Mandate for Change, 1953–1956* (New York: Doubleday, 1963), p. 301.

97 **Clark writes that.** Clark, *op. cit.*, p. 28.

97 **Senator H. Alexander Smith.** Cited in "Seaway Support Grows in Senate," *New York Times*, January 19, 1954.

CHAPTER SEVEN

98 **Robert Moses was.** *New York Times* obituary, July 31, 1981: www.nytimes.com/learning/general/onthisday/bday/1218.html [Retrieved 3/15/15].

99 **In his introduction.** Robert Caro, *The Power Broker: Robert Moses and the Fall of New York*, (New York: Knopf, 1972), p. 21.

99 **Moses in return.** "Robert Moses' Response to Robert Caro's *The Power Broker*," www.bridgeandtunnelclub.com/detritus/moses/ [Retrieved 03/19/15].

100 **An interest in political.** Caro, *op. cit.*, p. 112.

101 **This allowed, for example.** *Ibid.*, p. 616.

101 **In an especially telling.** *Ibid.*, p. 246.

101 **It is difficult.** Jill Murman-Payne, NYPA Timeline, December 1, 1954.

101 **Within a few years.** Daniel Macfarlane, *Negotiating a River: Canada, the US, and the Creation of the St. Lawrence Seaway* (British Columbia: UBC Press, 2014), p. 175.

102 **As general manager.** *New York Times* obituary, October 6, 1992.

102 **"On my insistence.** Robert Moses, *Public Works: A Dangerous Trade* (New York: McGraw-Hill, 1970), p. 340.

103 **There is the St. Lawrence.** Caro, *op. cit.*, pp. 8–9.

103 **"This is better than.** Cited in Kevin O'Keefe NYPA video, "St. Lawrence–FDR Power Project Construction," (at 2:65).

103 **Digging for the seaway.** *Ibid.*, pp. 132, 133.

103 **Although overall the safety.** *Ibid.*, p. 135.

104 **Sales to other neighboring.** Jill Murman-Payne, *op. cit.*

104 **For example, "the US government.** Macfarlane, *op. cit.*, p. 120.

104 **One Canadian expressed.** *Ibid.*, p. 121.

105 **Moses himself declared.** New York Power Authority, "Minutes of Trustee Meetings," May 16, 1955.

105 **The Queen unveiled.** Cited in *Bold Dream... Shining Legacy, Seventy-Fifth Anniversary of NYPA History* (NYPA, 2006).

106 **The American Society of Civil Engineers.** Macfarlane, *op. cit.*, pp. 135–6.

106 **One worker was killed.** Don Glynn, "The Collapse of Schoellkopf," *Niagara Gazette*, May 26, 2006.

107 **Then New York's former.** *Bold Dream... op. cit.*

108 **For design and compliance.** *Ibid.*; Ken Glennon, *Hard Hats of Niagara* (Indianapolis: Dog Ear, 2011), p. xvii.

108 **The Power Authority employed.** Glynn, *op. cit.*

109 **The conduit "seemed to.** Ken Glennon, *op. cit.*, pp. xv, 128.

110 **The diversion would reduce.** NYPA website: www.nypa.gov/facilities/niagara.htm [Retrieved 4/2/15].

110 **That is why the design.** "Niagara Falls History of Power," niagarafrontier.com [Retrieved 4/2/15].

111 **Moses planned to flood.** Edmund Wilson, "Apologies to the Iroquois," *The New Yorker*, October 17, 1959, p. 49, republished as *Apologies to the Iroquois and Their Neighbors* (Syracuse: Syracuse University, 1992), p. 139.

111 **According to Ginger Strand's.** Ginger Strand, *Inventing Niagara: Beauty, Power, and Lies*, (New York: Simon and Schuster, 2008), p. 184.

111 **"How our democratic system.** Wilson, *op. cit.*, p. 154.

111 **As the renowned.** *Ibid.*

112 **Moses escalated his threats.** *Ibid.*, p. 158.

112 **The Federal Court of Appeals.** Tuscarora Indian Nation v. Federal Power Commission, 265 F.2d 338 (D.C. Cir. 1959).

112 **Saving costs at the expense.** Cited in Edmund Wilson, *op. cit.*, p. 155.

112 **"I'm delighted,"** *Ibid.*, p. 157.

112 **Besides, the ruling noted.** Supreme Court ruling in *FPC v. Tuscarora Indian Nation* (1960) at: caselaw.lp.findlaw.com/scripts/getcase.pl?court=US&vol=362&invol=99 [Retrieved 4/10/15].

113 **Justice Hugo Black wrote.** *Ibid.*

113 **Strand writes that.** Strand, *op. cit.*, pp. 182–3.

113 **Despite the tragic deaths.** "Niagara Falls History of Power," *op. cit.*

114 **In a good year.** *Bold Dream..., op cit.*

114 **A plant near Niederwartha.** "Water Pumped Uphill to Keep Dynamos Busy," *Popular Science Monthly*, April 1930, p. 50.

115 **Then, whenever the state.** "A Ten-Mile Storage Battery," *Popular Science Monthly*, June 1930, p. 60.

115 **Seventy-five years later.** DOE Global Energy Storage Database: www.energystorageexchange.org/projects/261 [Retrieved 4/14/15].

115 **According to the Electric.** "99 Percent" cited in "Energy Storage: Packing Some Power," *The Economist*, March 3, 2012; International Energy Statistics, US Energy Information Administration at: www.eia.gov/cfapps/ipdbproject/iedindex3.cfm?tid=2&pid=82&aid=7&cid=regions&syid=2004&eyid=2010&unit=MK [Retrieved 4/14/15].

CHAPTER EIGHT

117 **"But the money came.** Robert Caro, *The Power Broker: Robert Moses and the Fall of New York*, (New York: Knopf, 1972), p. 1078.

118 **"The improvements made.** *New York Times*, February 15, 1988 [Retrieved 4/23/15].

119 **"Railroads halted."** *New York Times*, November 10, 1965, p. 1.

119 **Famously, nine months later.** *New York Times*, August 10, 1966; J. Richard Udry, *Demography*, August 1970.

119 **It is true, however.** David Frum, *How We Got Here: The '70s* (New York: Basic Books, 2000), p. 14.

119 **According to the Canadian.** Canadian Broadcasting Corporation, at: www.cbc.ca/archives/entry/1965-the-great-northeastern-blackout [Retrieved 4/21/15].

120 **One important outcome.** *Ibid.*; www.nerc.com [Retrieved 4/21/15]. It is now the North American Electric Reliability Corporation.

121 **The cause of that event.** "US-Canada Power System Outage Task Force Final Report" (2004) [Retrieved 7/31/15].

121 **According to one analyst.** Susan F. Tierney and Edward Kahn, "A Cost-Benefit Analysis of the New York Independent System Operator: The Initial Years" (Boston: Analysis Group, 2007), p. 9.

122 **The first nuclear power plant.** "From Obninsk Beyond: Nuclear Power Conference Looks to Future," IAEA, 24 June 2004, www.iaea.org/newscenter/news/obninsk-beyond-nuclear-power-conference-looks-future [Retrieved 6/5/15].

122 **At the end of 1957.** Energy Information Administration, *US Commercial Nuclear Power Historical Perspective, Current Status, and Outlook*, DOE/eia-0315 (Washington, D.C., March 1982), p. 10. *Annual Energy Review 1984*, p. 171.

122 **Until it didn't.** Energy Information Administration, "Fuel Choices in Steam Electric Generation: A Retrospective Analysis," Vol. 1, Overview, Draft Report, Table 2, cited in Rebecca A. McNerney, *Changing Structure of the Electric Power Industry: An Update* (Diane Publishing, 1998), p. 109.

123 **New York's Senator Robert.** *Bold Dream... Shining Legacy, Seventy-Fifth Anniversary of NYPA History* (NYPA, 2006).

123 **Just such a bill.** *Ibid.*

125 **In contrast to later.** *Ibid.*

126 **"More than any other**. *Ibid.*

127 **A few months later.** FitzPatrick at University of Vermont, May 3, 1977, cited in *Bold Dream... op. cit.*

127 **Looting broke out.** "The Blackout of 2003: The Past," Martin Gottlieb and James Glanz, *New York Times*, August 15, 2003 [Retrieved 7/10/15].

128 **"The looters were looting.** *Ibid.*

CHAPTER NINE

130 **That year the Swedish.** Svante Arrhenius, *Worlds in the Making: The Evolution of the Universe* (New York: Harper & Row, 1908), p. 61.

131 **That same year.** Paul R. Ehrlich, *The Population Bomb (*San Francisco: Sierra Club, 1968), p. 52.

131 **This first "oil shock".** "OPEC Oil Embargo 1973–1974," US Dept. of State, Office of the Historian: history.state.gov/milestones/1969-1976/OPEC [Retrieved 11/19/14.]

132 **According to the State.** *Ibid.*

132 **"The electric power industry.** Richard Rudolph and Scott Ridley, *Power Struggle: The Hundred-Year War over Electricity* (New York: Harper & Row, 1986), p. 197.

132 **Forty-five percent.** Cited in *Bold Dream... Shining Legacy, Seventy-Fifth Anniversary of NYPA History* (NYPA, 2006).

132 **"By September, prices.** Leonard S. Hyman, *America's Electric Utilities: Past, Present and Future* (Public Utilities Reports, Inc., Arlington, VA, 1997) p. 146.

133 **It was only natural.** NYPA timeline, at: www.nypa.gov/press/week/week.htm [Retrieved 12/14/14].

134 **As skilled foreign workers.** "Another Crisis for the Shah," *Time Magazine*, November 13, 1978 [Retrieved 12/4/14].

134 **Most US vehicles.** J. Leggett, *Half Gone: Oil, Gas, Hot Air and the Global Energy Crisis* (2005), p. 150.

135 **Eight years later.** US House of Representatives, Committee on Interior and Insular Affairs, Subcommittee on General, Oversight and Investigations (1987). "NRC Coziness with Industry: Nuclear Regulatory Commission Fails to Maintain Arms Length Relationship with the Nuclear Industry." An Investigative Report, 100th Congress First Session, Vol. 4 [Retrieved 12/11/14].

136 **And less than a month.** NYPA Press Release, January 18, 1980.

137 **The *New York Times* once.** "John Dyson, Upstart from Upstate, Speaks Up (Again)," *New York Times*, July 3, 1983 [Retrieved 12/4/14].

137 **As he declared.** Cited in *Bold Dream… op. cit.*

137 **The oil cartel's grip.** Dyson on WNBC's "Newscenter Forum," quoted in the *Evening News*, April 17, 1979.

138 **"First, I would say.** John Dyson interview with the author, 6/15/15.

138 **"When we got there,"** Telephone interview with James Cunningham by the author, 7/16/15.

138 **"We made quite an.** *Ibid.*

140 **Building upon a pilot.** PASNY "Button Up" Press Release, March 10, 1980.

141 **Button Up and similar.** *Ibid.*

142 **It seemed to me.** John Dyson interview *op. cit.*

143 **And not by accident.** James Cunningham interview *op. cit.*

143 **By 1981, NYPA sold more.** American Public Power Association survey results cited in NYPA Timeline, February 18, 1983.

146 **But the investor-owned.** John Dyson interview *op. cit.*

146 **The Public Service Commission.** "Fight Over New York Power Line Going to Court," *New York Times*, June 26, 1988.

147 **"By September 1982,"** "The Marcy-South Story," *NYPA Publications*, p 7.

148 **That simple strategy.** James Cunningham interview *op. cit.*

148 **By 1988, a group of.** "The Marcy-South Story," *op. cit.*, pp. 8–9.

148 **An editorial in.** *Ibid.*, p. 19.

149 **"It would mean economic.** Alex Radin, Letter to Editor, *New York Times*, November 4, 1983.

CHAPTER TEN

150 **"He was Governor Franklin.** This and the quotations that follow are from a Richard Flynn interview with the author, 7/8/15. Vyacheslav Molotov was the Soviet Union's Minister of Foreign Affairs.

151 **"During my time.** *NYPA Annual Report, 1976*, p. 2.

152 **Nonetheless, by May 1988.** *Bold Dream... Shining Legacy, Seventy-Fifth Anniversary of NYPA History* (NYPA, 2006).

153 **Construction began in May 1989.** Richard Flynn interview with the author, 7/8/15.

153 **Government is the coming.** Governor Mario Cuomo, Long Island Sound Cable Ceremony, Garden City, Long Island, May 30, 1991. [Emphasis added.] In 1995, a new contract for purchases from Hydro-Québec was to take effect.

155 **In 1990 we won.** Richard Flynn interview *op. cit.*

156 **Meanwhile, the projected.** This account owes largely to Don Fagin, "Lights Out at Shoreham," *Newsday*, November 3, 1007, from web.archive.org/web/20071201005429/www.newsday.com/community/guide/lihistory/ny-history-hs9shore,0,563942.story [Retrieved 7/26/15].

156 **That was how its.** Kenneth F. MacCallion, *Shoreham and the Rise and Fall of the Nuclear Power Industry* (1995), p. 11.

156 **Community organizers continued.** *Ibid.*, p. 7.

157 **Then came the nuclear.** John T. McQuiston, "Shoreham Action is One of Largest Held Worldwide," *New York Times*, June 4, 1979.

157 **Eight months later local.** Nora Bredes obituary, *New York Times*, August 22, 2011.

157 **Dozens of groups.** Fagin, *op. cit.*

158 **Originally Shoreham came.** Richard Flynn interview *op. cit.*

159 **"We were using energy.** Steve Lerner, *Eco-Pioneers: Practical Visionaries Solving Today's Environmental Problems* (Cambridge, MA: MIT Press, 1998), p. 92.

159 **But Freeman was also.** S. David Freeman interview in "Conversations with History" (Institute of International Studies, UC Berkeley, 2003): globetrotter.berkeley.edu/people3/Freeman/freeman-con0.html [Retrieved 8/6/15].

159 **He started another program.** Steve Lerner, *op. cit.*, p. 98.

160 **Perhaps the best symbol.** *Ibid.*, p. 93.

160 **According to one environmental.** *Ibid.*, p. 91.

160 **In New York, Freeman.** *Bold Dream... op. cit.*

160 **"From a straight business.** Matthew Wald, "State Power Authority Chief Opposes Contract With Quebec," *New York Times*, March 3, 1994 [Retrieved 7/29/15].

161 **For his various efforts.** Jill Murman-Payne, NYPA Timeline, December 1, 1954.

161 **In July 1995.** "State Power Agency to Name Leader," *New York Times*, July 25, 1995.

161 **Thanks to his "personable.** *Bold Dream... op. cit.*

162 **Congress had tried.** Leonard S. Hyman, *America's Electric Utilities: Past, Present and Future* (Public Utilities Reports, Inc., Arlington, VA, 1997), p. 152.

164 **"Competition works best.** Cited in *Bold Dream... op. cit.*

164–5 **Eugene W. Zeltmann.** *Ibid.*

165 **Rappleyea, speaking at the ceremonial.** *Ibid.*

166 **As Governor Pataki noted.** "Power Authority Headquarters Named in Honor of Chairman Rappleyea," NYPA Press Release, January 30, 2001.

CHAPTER ELEVEN

169 **NYCHA's stated mission.** *NYCHA Fact Sheet*, www.nyc.gov/html/nycha/html/about/factsheet.shtml [Retrieved 8/18/14].

172 **Perhaps most significantly.** Using EPA calculator at www.epa.gov/cleanenergy/energy-resources/calculator.html.

172 **As a result of all.** www.queensbuzz.com/ny-power-authority-deconstructs-poletti-power-plant-cms-994 [Retrieved 9/28/14].

173 **At the end of 2012.** www.governor.ny.gov/executiveorder/88 [Retrieved 8/30/14].

173 **More than 90 percent.** BuildSmart website, www.buildsmart.ny.gov/about/ [Retrieved 8/30/14].

174 **"From an economic perspective.** www.nypa.gov/BuildSmartNY/BaselineEnergyReport08-2013.pdf [Retrieved 10/9/14].

175 **Local environmental groups.** Much of this account is reported from www.queensbuzz.com/ny-power-authority-deconstructs-poletti-power-plant-cms-994 [Retrieved 9/28/14].

179 **These actions, along with.** *NYPA Sustainability Report 2012.*

180 **In all, NYPA has helped.** www.nypa.gov/ev/ [Retrieved 12/9/15].

181 **Across the country.** American Public Power Association website, www.publicpower.org [Retrieved 9/23/14].

182 **It is only thanks.** *Ibid.*

183 **According to the New.** www.nyiso.com/public/about_nyiso/understanding_the_markets/benefits_of_markets/index.jsp [Retrieved 6/2/15].

184 **But in 2014.** www.whitehouse.gov/sites/default/files/image/president27 sclimateactionplan.pdf [Retrieved 9/25/14].

184 **Already, in Copenhagen.** *New York Times*, Nov. 29, 2009.

CHAPTER TWELVE

186 **Less than 60 years.** www.scrippsco2.ucsd.edu/ [Retrieved 10/10/14].

186 **According to both NOAA's.** NOAA website: www.ncdc.noaa.gov/sotc/global/ [Retrieved 10/06/14].

186 **Also since 1880.** www2.ucar.edu/climate/faq/how-much-has-global-temperature-risen-last-100-years; www.ncdc.noaa.gov/sotc/global/201513 [Retrieved 10/14/16].

187 **Globally, CO_2 emissions.** Global Carbon Project at www.globalcarbonproject.org [Retrieved 10/10/14].

187 **According to University of Chicago.** D. Archer, *Long Thaw: How Humans Are Changing the Next 100,000 Years of Earth's Climate* (Princeton Univ. Press, 2008), cited in *Nature*, September 12, 2008. www.nature.com/climate/2008/0812/full/climate.2008.122.html [Retrieved 10/10/14].

187 **Although globally one-third.** www2.epa.gov/sites/production/files/2013-09/documents/20130920factsheet.pdf [Retrieved 10/10/14].

187 **According to the Environmental.** Reported by the Associated Press, January 11, 2012.

188 **No such explanation.** Ross Gelbspan, *The Heat Is On: The Climate Crisis, The Cover-Up, The Prescription* (Basic Books, 1998).

188 **In late August 2011.** Lixion A. Avila and John Cangialosi, "Hurricane Irene Tropical Cyclone Report," December 14, 2011, at www.nhc.noaa.gov/data/tcr/AL092011_Irene.pdf [Retrieved 10/19/14].

190 **Nationally, 286 people.** US DOE, www.oe.netl.doe.gov/docs/2012_SitRep6_Sandy_10312012_1000AM_v_1.pdf [Retrieved 10/17/14].

190 **When "Superstorm Sandy".** CNN at www.cnn.com/2013/07/13/world/americas/hurricane-sandy-fast-facts/ [Retrieved 10/22/14].

190 **Six subway stops.** www.businessinsider.com/heres-how-nycs-subway-sytem-has-come-back-from-hurricane-sandy-2013-10 [Retrieved 10/22/14].

190 **Then came an explosion.** Gov. Andrew Cuomo, "We Will Lead on Climate Change," *New York Daily News*, November 15, 2012.

191 **So as the staff began.** FEMA Report, www.fema.gov/disaster/4085/updates/new-yorks-bellevue-hospital-takes-mitigation-steps-after-hurricane-sandy [Retrieved 10/22/14].

191 **"We were walking.** In private conversation with the author, 11/22/12.

191 **These included forty-four.** FEMA statistics, according to the *Huffington Post,* www.huffingtonpost.com/tag/hurricane-sandy-long-island/ [Retrieved 10/22/14].

191 **Across the state.** For photographs of the damage in New York and New Jersey, see *The Atlantic*, November 2012, www.theatlantic.com/infocus/2012/11/hurricane-sandy-the-aftermath/100397/ [Retrieved 10/23/14].

191 **Staten Island's coast.** Irving DeJohn, "Hurricane Sandy, one year later," *NY Daily News*, October 26, 2013.

192 **The rationing continued.** FEMA timeline at www.fema.gov/hurri cane-sandy-timeline and Climate Central at www.climatecentral.org /news/ongoing-coverage-of-historic-hurricane-sandy-15184 [Retrieved 10/23/14].

193 **Extreme weather is.** Governor Andrew Cuomo, "We Will Lead on Climate Change," *New York Daily News*, November 15, 2012.

195 **California leads the large.** California Smart Grid Roadmap at: www.caiso.com /Documents/SmartGridRoadmapandArchitecture.pdf [Retrieved 10/25/14].

195 **This facility was built.** www.governor.ny.gov/press/10212014-energy-management-network [Retrieved 10/24/14].

195 **In 2013, the United States** Giles Parkinson, "Solar Grid Parity In All 50 US States By 2016, Predicts Deutsche Bank," *Clean Technica*, October 24, 2014 [Retrieved 12/14/15].

195 **As of 2014.** Vishal Shah, Deutsche Bank analyst, quoted in www.tree hugger.com/renewable-energy/rooftop-solar-power-will-be-grid-parity-50-us-states-2016-says-deutsche-bank.html [Retrieved 10/30/14].

196 **"Electricity is generated.** *Reforming The Energy Vision*, NYS Dept. of Public Service, Staff Report and Proposal, Case 14-M-0101, April 24, 2014.

196 ***Reforming the Energy Vision.*** *Ibid.*, p. 1.

198 **From NYPA's perspective.** microgridknowledge.com/nypa-pursues-microgrids-re-invents-mission-new-grid/ [Retrieved 10/29/14].

198 **"In the years between.** David Nye, *Electrifying America: Social Meanings of a New Technology* (MIT Press, 1990), p. 138.

199 **"New York," he declares.** Governor Andrew Cuomo, cited in *New York Times*, September 10, 2014.

Bibliography

Ackerman, Kenneth D. *Boss Tweed: The Rise and Fall of the Corrupt Pol Who Conceived the Soul of Modern New York* (New York: Carroll & Graf, 2005).

Adams, Edward Dean, *Niagara Power*, 2 vols., (Niagara Falls: Niagara Falls, Power Co., 1927).

Adams, Henry, "The Dynamo and the Virgin" in *The Education of Henry Adams* (New York: Random House, 1931).

Archer, David, *The Long Thaw: How Humans Are Changing the Next 100,000 Years of Earth's Climate* (Princeton: Princeton University Press, 2008).

Arrhenius, Svante, *Worlds in the Making: The Evolution of the Universe* (New York: Harper & Row, 1908).

Bauer, John and Costello, Peter, *Public Organization of Electric Power* (New York: Harper and Bros., 1949).

Berton, Pierre, *Niagara: A History of the Falls* (New York: Penguin, 1992).

Brock, Barbara H., *The Development of Public Utility Accounting in New York* (East Lansing: Michigan State University, 1981).

Burrows, Edwin G. and Wallace, Mike, *Gotham: A History of New York City to 1898* (New York: Oxford Paperbacks, 2000).

Caro, Robert, *The Power Broker: Robert Moses and the Fall of New York* (New York: Vintage, 1974).

Cheney, Margaret, *Tesla: Man Out of Time* (New York: Simon and Schuster, 2011).

Clare, W.H., *The Case of Public Ownership of Electric Utilities* (Chicago: Public Ownership League of America, 1937).

Clark, Austin W., "Fish or Cut Bait? Dwight D. Eisenhower and the Creation of the St. Lawrence Seaway" (unpublished thesis, Gettysburg College, 2012).

Cromwell, Thomas, *Municipal Monopolies* (New York: Edward Bemis, 1899).

Dickens, Charles, *Life, Letters and Speeches of Charles Dickens* (New York: Houghton, Mifflin, 1894).

Dow, Charles Mason, ed., *Anthology and Bibliography of Niagara Falls*, 2 vols. (Albany: State of New York).

Energy Information Administration, *Annual Outlook for U.S. Electric Power, 1985* (Washington, D.C.: EIA, 1985).

Evans, Harold, *They Made America: From the Steam Engine to the Search Engine, Two Centuries of Innovators* (Boston: Back Bay Books, 2004).

Ehrlich, Paul R., *The Population Bomb* (San Francisco: Sierra Club, 1968).

Essig, Mark, *Edison and the Electric Chair: A Story of Light and Death* (New York: Bloomsbury, 2006).

Frum, David, *How We Got Here: The '70s* (New York: Basic Books, 2000).

Gelbspan, Ross, *The Heat Is On: The Climate Crisis, The Cover-Up, The Prescription* (New York: Basic Books, 1998).

Glennon, Ken, *Hard Hats of Niagara* (Indianapolis: Dog Ear, 2011).

Heilbron, J. L., *Electricity in the 17th and 18th Century* (Berkeley: University of California Press, 1979).

Hubbard, Preston John, *Origins of the TVA* (Nashville: Vanderbilt University, 1961).

Hughes, Charles, E., *Public Papers of Charles E. Hughes, 1907* (New York: J.B. Lyon, 1909).

Hughes, Thomas P., *Networks of Power: Electrification in Western Society, 1880-1930* (Baltimore: Johns Hopkins University Press, 1983).

Hughes, Thomas P., "Technological History and Technical Problems," in Chauncey Starr and Philip C. Ritterbush, eds., *Science, Technology and the Human Prospect* (New York: Pergamon Press, 1980).

Hyman, Leonard S., *America's Electric Utilities: Past, Present and Future* (Public Utilities Reports, Inc., Arlington, VA, 1997).

Israel, Paul, *Edison: A Life of Invention* (New York: Wiley and Sons, 1998).

Jackson, Kenneth T., *The Encyclopedia of New York City* (New Haven, Connecticut: Yale, 1995).

Jonnes, Jill, *Empires of Light: Edison, Tesla, Westinghouse and the Race to Electrify The World* (New York: Random House, 2003).

Josephson Matthew, *Edison: A Biography* (New York: McGraw-Hill, 1959).

Kilroe, Frank E., "The Governorship of Charles Evans Hughes: A Study in Reform (1906-1910)" (unpublished M.A. Thesis, Columbia University, 1934).

King, Judson, *The Conservation Fight: From Theodore Roosevelt to the Tennessee Valley Authority* (Washington, D.C.: Public Affairs, 1959).

Klein, Maury, *The Power Makers: Steam, Electricity, and the Men Who Invented Modern America* (New York: Bloomsbury, 2010).

Landes, David. S. *The Unbound Prometheus: Technological Change and Industrial Development in Western Europe from 1750 to the Present* (New York: Cambridge University Press, 1969).

Leggett, Jeremy, *Half Gone: Oil, Gas, Hot Air and the Global Energy Crisis* (London: Portobello Books, 2005).

Lerner, Steve, *Eco-Pioneers: Practical Visionaries Solving Today's Environmental Problems* (Cambridge: MIT Press, 1998).

Leupp, Francis Ellington, *George Westinghouse: His Life and Achievements* (New York: Little, Brown, 1918).

Lindley, Ernest K., *Franklin D. Roosevelt: A Career in Progressive Democracy* (Indianapolis: Bobbs-Merrill, 1931).

Macfarlane, Daniel, *Negotiating a River: Canada, the US, and the Creation of the St. Lawrence Seaway* (British Columbia: UBC Press, 2014).

McChesney, Robert, and Podesta, John, "Let There Be Wi-Fi" (*Washington Monthly*, Jan./Feb. 2006).

McDonald, Forrest, *Insull* (Chicago: University of Chicago Press, 1962).

McNichol, Tom, *AC/DC: The Savage Tale of the First Standards War* (New York: Jossey-Bass, 2013).

Mitgang, Herbert. *The Man Who Rode the Tiger: The Life and Times of Judge Samuel Seabury* (Philadelphia: Lippincott, 1963).

Morgan, Richard, Riesenberg, Tom, and Troutman, Michael, *Taking Charge: A New Look at Public Power* (Washington, D.C.: Environmental Action Foundation, 1976).

Morris, Edmund. *The Rise of Theodore Roosevelt* (New York: Random House, 2012).

Moses, Robert, *Public Works: A Dangerous Trade* (New York: McGraw-Hill, 1970).

Nye, David E., *Electrifying America: Social Meanings of a New Technology* (Cambridge: MIT Press, 1990).

O'Neill, John J., *Prodigal Genius: The Life of Nikola Tesla* (New York: McKay, 1944).

Passer, Harold C., *The Electrical Manufacturers, 1875-1900: A Study in Competition, Entrepreneurship, Technical Chane, and Economic Growth* (Cambridge: Harvard, 1953).

Phillips, Sarah, "FDR, Hoover, and the New Rural Conservation," in Henderson, Henry L., and Woolner, David B., eds., *FDR and the Environment* (New York: Palgrave Macmillan, 2009).

Pope, Leonard Franklin, *Evolution of the Electric Incandescent Lamp* (Los Angeles: University of California, 1889).

Prout, Henry G., *A Life of George Westinghouse* (Ann Arbor: University of Michigan, 1926).

Pusey, Merlo J., *Charles Evans Hughes* (New York: Macmillan, 1951).

Raushenbush, H.S., and Laidler, Harry W., *Power Control* (New York: New Republic Inc., 1928).

Rudolph, Richard, and Ridley, Scott, *Power Struggle: The Hundred-Year War Over Electricity* (New York: HarperCollins, 1986).

Schlesinger, Arthur M., *The Age of Roosevelt: The Crisis of the Old Order* (Boston: Houghton Mifflin, 1957).

Seifer, Marc J., *Wizard: The Life and Times of Nikola Tesla* (New York: Citadel, 1996).

Skrabec, Quentin R., Jr., *George Westinghouse: Gentle Genius* (New York: Algora Publishing, 2006).

Spieler, Cliff, and Hewitt, Tom, *Niagara Power from Joncaire to Moses* (Lewiston, NY: Niagara Power Publishers, 1959).

Strand, Ginger. *Inventing Niagara: Beauty, Power, and Lies* (New York: Simon and Schuster, 2008).

Thomas, John M., *Michael Faraday and the Royal Institution* (New York: Adam Hilger, 1991).

Thompson, Carl D., *Municipal Electric Light and Power Plants in the United States and Canada* (Chicago: Public Ownership League of America, 1922).

Tierney, Susan F., and Kahn, Edward, *A Cost-Benefit Analysis of the New York Independent System Operator: The Initial Years* (Boston: Analysis Group, 2007).

Tollefson, Cene, *Bonneville Power Administration and the Struggle for Power at Cost* (Portland: Bonneville Power Administration, 1987).

Wells, H. G., "The End of Niagara," *The Future in America: A Search After Realities* (New York: Harper and Bros., 1906).

Wesser, Robert E., *Charles Evans Hughes: Politics and Reform in New York, 1905-1910* (Ithaca: Cornell University Press, 1967).

Zacks Richard, *Island of Vice: Theodore Roosevelt's Doomed Quest to Clean Up Sin-Loving New York* (New York: Doubleday, 2012).

Index

About the Author

Dr. Rock Brynner is a novelist and historian, author of *The Doomsday Report* (1998), *Dark Remedy: The Impact of Thalidomide and its Revival as a Vital Medicine* (2001) and *Empire and Odyssey: The Brynners in Far East Russia and Beyond* (2006). He earned his M.A. in Philosophy at Trinity College, Dublin, and his M.A. and Ph.D. in History at Columbia University. He taught US History at Marist College and was Assistant Professor of Political Science at Western Connecticut State University. Since 2012, he has served on the New York State Task Force on Life and the Law.

COSIMO

CPSIA information can be obtained
at www.ICGtesting.com
Printed in the USA
BVOW10s1948230817
492922BV00004B/11/P